AF445474

Novel Bioderived Composites from Wastes

Novel Bioderived Composites from Wastes

Editors

Andrea Petrella
Marco Race
Danilo Spasiano

MDPI • Basel • Beijing • Wuhan • Barcelona • Belgrade • Manchester • Tokyo • Cluj • Tianjin

Editors
Andrea Petrella
Polytechnic University of Bari
Italy

Marco Race
University of Cassino and Southern Lazio
Italy

Danilo Spasiano
Polytechnic University of Bari
Italy

Editorial Office
MDPI
St. Alban-Anlage 66
4052 Basel, Switzerland

This is a reprint of articles from the Special Issue published online in the open access journal *Materials* (ISSN 1996-1944) (available at: https://www.mdpi.com/journal/materials/special_issues/Novel_Bioderived_Composites_Wastes).

For citation purposes, cite each article independently as indicated on the article page online and as indicated below:

LastName, A.A.; LastName, B.B.; LastName, C.C. Article Title. *Journal Name* **Year**, *Article Number*, Page Range.

ISBN 978-3-03943-110-6 (Pbk)
ISBN 978-3-03943-111-3 (PDF)

Contents

About the Editors

Andrea Petrella, an assistant professor, was born in Bari in 1976. He graduated with an MSc in Chemistry at the University of Bari in July 2001, and gained his Ph.D. in Chemical Sciences in 2005 from the University of Bari. From 2007, he has been an assistant professor of Materials Science and Technology at the Polytechnic University of Bari. His research fields are summarized as follows: use of recycling organic and inorganic materials in the building trade and/or in the removal of heavy metals present in wastewater; photocatalytic materials for the degradation of bio-persistent pollutants in water and wastewater; nanocomposites for energy conversion and for novel optical devices. He has co-authored more than 50 papers published in international journals and indexed on Scopus.

Marco Race, an assistant professor, was born in Napoli; he graduated with an MSc in Environmental Engineering at Università degli Studi di Napoli Federico II in May 2012, and gained his Ph.D. in Environmental Systems Analysis at UNINA in 2016, according to the European Label. From 2018, he has been an assistant professor at University of Cassino. His main research fields concern the treatment of waste or wastewater treatment, the remediation of soil and groundwater, novel contaminant (bio)monitoring and risk assessment approaches, and trace metals and organics in biogeochemical cycles. He is the author of more than 50 papers published in international journals, conference proceedings and books chapters. He gained the international award on soil reclamation.

Danilo Spasiano, Assistant Professor, was born in Naples in 1984. He graduated as MSc in Environmental Engineering at Università degli Studi di Napoli Federico II in October 2009, and obtained his Ph.D. in Chemical Engineering at the same university in 2013. From 2015, Danilo Spasiano has been an assistant professor in Sanitary Engineering at Polytechnic of Bari. His main research fields concern the treatment of asbestos, containing wastes and wastewater treatment by means of advanced oxidation process. He has coauthored more than 40 papers published in international journals and indexed on Scopus.

Editorial

Novel Bioderived Composites from Wastes

Andrea Petrella [1,*], Marco Race [2] and Danilo Spasiano [1]

[1]   Department of Civil, Environmental, Land, Building Engineering and Chemistry, Polytechnic University of Bari, via E. Orabona, 4, 70125 Bari, Italy; danilo.spasiano@poliba.it

[2]   Department of Civil and Mechanical Engineering, University of Cassino and Southern Lazio, Via di Biasio 43, 03043 Cassino, Italy; marco.race@unicas.it

*   Correspondence: andrea.petrella@poliba.it; Tel.: +39-(0)8-0596-3275; Fax: +39-(0)8-0596-3635

Received: 28 May 2020; Accepted: 29 May 2020; Published: 5 June 2020

1. Introduction

The recycling and reuse of solid wastes can be considered important challenges for civil and environmental applications in the frame of a more sustainable model of development and the consumption of new resources and energy [1–5]. The recovery of raw materials from nonconventional sources and their transformation into usable resources not only represents an economic advantage, but also offers an ecological opportunity for the utilization of by-products which would otherwise be landfilled [6–11]. In this respect, these secondary raw materials, generally derived from industrial, agricultural and food manufacturing activities, become an abundant resource that can be easily reused for different applications, as reported in the recent studies collected in this Special Issue.

For that purpose, six papers were related to the preparation of innovative composite materials. Specifically, five papers reported the reuse of end-of-life tire rubber, porous glass, expanded polystyrene, slags, fly ashes and sheep's wool fibers for the preparation of cement conglomerates [12–16], while the last one reported the reuse of amorphous silica nanoparticles for the preparation of composites with natural rubber [17]. Moreover, five papers were related to the treatment of wastes for environmental applications. Specifically, two papers reported the reuse of egg by-products [18] and crab shell [19] for the removal of biopersistent micropollutants, and two papers reported the reuse of white bamboo fibrils as oil absorbent [20] and a rapid method for the disposal of radioactive contaminated soil waste [21], respectively. The last one reported the use of wheat straw biochar for cobalt sorption from contaminated soil [22].

2. Waste Products for Construction Materials

In the papers by Petrella and coworkers [12,13], recycled materials, such as end-of-life tire rubber (TR), porous glass (PG) and expanded polystyrene (EPS) were used as aggregates for the production of unconventional cement mortars. A cheap and environmentally safe process was employed, since no pre-treatment of the renewable aggregates was carried out. The thermal conductivity of these lightweight composites was 80–90% lower than the conventional sand mortars. Moreover, the presence of the recycled glass (PG) influenced the mechanical strengths and the thermal insulation of the specimens, thanks to the high stiffness and closed porosity of the aggregate [23], while the conglomerates with end-of-life tire rubber and expanded polystyrene showed thermal insulation and hydrophobic behavior due to the low water absorption. These results revealed that these composites may be suitable for nonstructural thermo-insulating products, specifically for the production of inside and outside elements.

Perez-Garcia and coworkers studied the properties of green cementitious grout mixtures characterized by cement substitution with slag (25% and 50%) derived from steel manufacturing [14]. The addition was carried out without additives and the slag was introduced as a cement replacement. Specifically, different slags (ladle furnace slag (LFS) and blast furnace slag (GGBS)) were used for the preparation of cheap conglomerates, which were tested for exudation, compressive and flexural strength, in order to analyze the feasibility of the mixtures for industrial applications. In general, these conglomerates showed a lower density and an improvement in fluency and viscosity with respect to the conventional references, while the mechanical response was dependent on the origin of the slag. The fluidity of the mixtures allows for their use in applications such as jet grouting or ground improvements. GGBS slags improved the mechanical strengths and workability of the mixtures, while the LFS slags can be employed in other types of works where a high strength is not required.

The paper authored by Wei-Ting Lin and coworkers [15] showed the feasibility of using ground-granulated blast-furnace slag (S) and circulating fluidized bed co-fired fly ash (FA) as non-cement binding materials. In fact, they determined the optimal mix proportions (100% cement replacement, S:FA ratios of 4:6, 5:5, 6:4, water/binder ratio of 0.55) in order to achieve high dimensional stability and good mechanical properties with inclusion in the resulting composite of polypropylene fibers. Composites with an S:FA ratio of 6:4 showed a compressive strength approximately equal to 30 MPa, which is 80% the strength of conventional cement-based materials at 28 days of curing. The strong influence of the polypropylene fibers in reinforcing the non-cement blended materials was demonstrated, indeed the inclusion of 0.2% fibers in the mixture further increased the compressive strength to 35 MPa and also enhanced the compactness of the micropore structures, increased the tensile strength and decreased absorption and the likelihood of shrinkage.

In the paper by Maia Pederneiras and coworkers [16] sheep's wool fibers were incorporated into mortars to ensure the durability of the render and improve the flexural strength, fracture toughness and impact resistance. The novel composites were prepared with cement and cement–lime ligands. The addition of 10% and 20% (in volume) of 1.5 cm and 3.0 cm wool fibers led to the increase in the ductility of the mortars and an improvement in the mechanical properties. In fact, these specimens showed high ductility because they presented a higher flexural and compressive strengths ratio ($\sigma f / \sigma c$) with respect to the reference mortars. The conglomerates also showed an improvement in the fracture toughness, with specific reference when longer fibers were incorporated. Moreover, the presence of longer fibers affected the increase in the flexural and compressive strengths. The wool fiber composites also presented a lower tendency to crack when compared with the conventional artifacts.

Nguyen and coworkers reported a method of recovering amorphous silica nanoparticles (40–60 nm) from hexafluorosilicic acid waste (Vietnamese fertilizer industry) through a precipitation process [17]. These particles were adopted as a reinforcing filler of natural rubber (NR) materials, which were characterized by morphological, mechanical, rheological and thermal measurements. Specifically, the mechanical properties of nanosilica-filled NR composites reached the optimum with 3 phr of nanosilica, and accordingly the tensile strength, hardness and decomposition temperature of these novel materials showed an improvement of 20.6%, 7.1%, and 2.5%, respectively, with respect to the pristine NR. The hardness of the filled samples increased with increasing nanosilica content, as opposed to the elongation at break. The improved mechanical properties can be explained by the tensile fractured surface morphology, which shows that the silica-filled NR is rougher than the pristine natural rubber sample.

3. Waste Materials for Environmental Science

A reusable adsorbent constituted by eggshell was proposed by Murcia-Salvador and coauthors for the removal of Direct Blue 78 (DB78) dye from wastewater [18]. Notably, the maximum adsorption of DB78 onto eggshell was obtained at pH 5 and 12.5 g/L of adsorbent dosage and the adsorption capacity of DB78 was 13 mg/g. The study of the thermodynamic parameters highlighted that the adsorption process was endothermic and spontaneous in the 29–75 °C range. In addition, the combination of the adsorption process on eggshell and the H_2O_2/pulsed light advanced oxidation process led to a further decrease in the pollutant concentration, thus demonstrating that the adoption of both processes can be used successfully in the removal of dyes at higher concentrations from wastewater.

The paper contributed by Rizzi and coworkers introduced the use of chitosan from biowaste (crab shell) to induce the formation of solid films useful for the decontamination of water from emerging pollutants [19]. In particular, ketoprofen was used as a contaminant, and a high percentage of removal, at least 90%, was shown in a short time under the proposed experimental conditions. Moreover, the authors detailed the nature of the adsorption by changing the chemical and physical parameters, such as the pH, temperature changes and electrolyte presence in the solutions containing the pollutant. The interaction between the ketoprofen carboxylic moiety and the chitosan amino groups were proposed by showing that the presence of salts inhibited the adsorption process, giving the opportunity to desorb the pollutant and recycle both the adsorbent material and ketoprofen.

Nguyen and coauthors reported the synthesis of highly porous cellulose aerogels, produced from white bamboo fibrils, which could be adopted to clean up oil spills and toxic chemicals in aquatic environments [20]. Specifically, white bamboo was cut and placed into an autoclave for 60 min. Afterwards, samples were immersed in a 2% NaOH solution at 70 °C to remove the cell walls and the obtained fibers were ground until they reached a micron-sized diameter. These cellulose fibers were dispersed in a $NaOH/urea/H_2O$ mixture, leading to a cellulose hydrogel, which was washed with water and then freeze-dried. Finally, MEMO silane was deposited on the cellulose-based aerogel. This silane-treated cellulose aerogel exhibited high absorption capacities of 1091 ± 19.6%, 1237 ± 17.6% and 1247 ± 21.1% by weight gain for waste motor oil, diesel and gasoline, respectively.

Xue and coworkers described a rapid and effective method for the disposal of radioactive contaminated soil waste [21]. For this purpose, simulated Ce-bearing radioactive soil waste was immobilized by the self-propagating high-temperature synthesis (SHS) of forms containing 5–25% of contaminated material and which were characterized by the analysis of phase composition, microstructure and chemical durability. The simulated nuclide Ce was immobilized into a pyrochlore-rich waste matrix characterized by multiphase composite materials (SiO_2, $Gd_2Ti_2O_7$ and Cu). Moreover, it was observed that the simulated nuclide Ce was simultaneously present in the pyrochlore and soil phases, thus indicating a partial migration of Ce during the SHS reaction. The solidified body of a Cu-20 sample (with 20% of soil waste) exhibited high stability.

Finally, Medyńska-Juraszek and coworkers used wheat straw biochar for cobalt sorption from contaminated soil [22]. It was demonstrated that this material was an efficient sorbent, decreasing the mobility and availability of Co^{2+} in soil and reducing health risks related to human exposure. The dominant mechanisms of sorption were mainly associated with interactions with carboxylic and hydroxyl groups present on the biochar surface. Cobalt immobilization was more complex because the efficiency of the process can be modified by biochar oxidation and interaction with soil constituents.

4. Outlooks

The above-mentioned papers have demonstrated that the recovery of solid wastes from industrial, agricultural and food manufacturing activities can be considered an important challenge for the design of new materials and for the evolution of new techniques in the frame of a more sustainable model of development and the consumption of new resources and energy.

Acknowledgments: We would like to thank all the authors and the reviewers. Special acknowledgments to Emma Fang and all the staff of the Materials Editorial Office for the great support during the preparation of this Special Issue.

Conflicts of Interest: The authors declare no conflict of interest.

References

1. Ferraro, A.; Dottorini, G.; Massini, G.; Miritana, V.M.; Signorini, A.; Lembo, G.; Fabbricino, M. Combined bioaugmentation with anaerobic ruminal fungi and fermentative bacteria to enhance biogas production from wheat straw and mushroom spent straw. *Bioresour. Technol.* **2018**, *260*, 364–373. [CrossRef]
2. Sengupta, A.; Gupta, N.K. MWCNTs based sorbents for nuclear waste management: A review. *J. Environ. Chem. Eng.* **2017**, *5*, 5099–5114. [CrossRef]

3. Li, M.; Liu, J.; Han, W. Recycling and management of waste lead-acid batteries: A mini-review. *Waste Manag. Res.* **2016**, *34*, 298–306. [CrossRef] [PubMed]

4. Petrella, A.; Petrella, M.; Boghetich, G.; Petruzzelli, D.; Calabrese, D.; Stefanizzi, P.; De Napoli, D.; Guastamacchia, M. Recycled waste glass as aggregate for lightweight concrete. *Proc. Inst. Civ. Eng. Constr. Mater.* **2007**, *160*, 165–170. [CrossRef]

5. Petrella, A.; Petrella, M.; Boghetich, G.; Basile, T.; Petruzzelli, V.; Petruzzelli, D. Heavy metals retention on recycled waste glass from solid wastes sorting operations: A comparative study among different metal species. *Ind. Eng. Chem. Res.* **2012**, *51*, 119–125. [CrossRef]

6. Ossa, A.; García, J.L.; Botero, E. Use of recycled construction and demolition waste (CDW) aggregates: A sustainable alternative for the pavement construction industry. *J. Clean. Prod.* **2016**, *135*, 379–386. [CrossRef]

7. Gómez-Meijide, B.; Pérez, I.; Pasandín, A.R. Recycled construction and demolition waste in cold asphalt mixtures: Evolutionary properties. *J. Clean. Prod.* **2016**, *112*, 588–598. [CrossRef]

8. Petrella, A.; Petruzzelli, V.; Basile, T.; Petrella, M.; Boghetich, G.; Petruzzelli, D. Recycled porous glass from municipal/industrial solid wastes sorting operations as a lead ion sorbent from wastewaters. *React. Funct. Polym.* **2010**, *70*, 203–209. [CrossRef]

9. Petrella, A.; Spasiano, D.; Rizzi, V.; Cosma, P.; Race, M.; De Vietro, N. Lead ion sorption by perlite and reuse of the exhausted material in the construction field. *Appl. Sci.* **2018**, *8*, 1882. [CrossRef]

10. Spasiano, D.; Luongo, V.; Petrella, A.; Alfè, M.; Pirozzi, F.; Fratino, U.; Piccinni, A.F. Preliminary study on the adoption of dark fermentation as pretreatment for a sustainable hydrothermal denaturation of cement-asbestos composites. *J. Clean. Prod.* **2017**, *166*, 172–180. [CrossRef]

11. Ferraro, A.; Farina, I.; Race, M.; Colangelo, F.; Cioffi, R.; Fabbricino, M. Pre-treatments of MSWI fly-ashes: A comprehensive review to determine optimal conditions for their reuse and/or environmentally sustainable disposal. *Rev. Environ. Sci. Bio/Technol.* **2019**, *18*, 453–471. [CrossRef]

12. Petrella, A.; Di Mundo, R.; De Gisi, S.; Todaro, F.; Labianca, C.; Notarnicola, M. Environmentally sustainable cement composites based on end-of-life tyre rubber and recycled waste porous glass. *Materials* **2019**, *12*, 3289. [CrossRef]

13. Petrella, A.; Di Mundo, R.; Notarnicola, M. Recycled expanded polystyrene as lightweight aggregate for environmentally sustainable cement conglomerates. *Materials* **2020**, *13*, 988. [CrossRef] [PubMed]

14. Perez-Garcia, F.; Parron-Rubio, M.E.; Garcia-Manrique, J.M.; Rubio-Cintas, M.D. Study of the suitability of different types of slag and its influence on the quality of green grouts obtained by partial replacement of cement. *Materials* **2019**, *12*, 1166. [CrossRef] [PubMed]

15. Lin, W.T.; Lin, K.L.; Korniejenko, K.; Fiala, L.; Cheng, A.; Chen, J. Composite properties of non-cement blended fiber composites without alkali activator. *Materials* **2020**, *13*, 1443. [CrossRef] [PubMed]

16. Maia Pederneiras, C.; Veiga, R.; de Brito, J. Rendering mortars reinforced with natural sheep's wool fibers. *Materials* **2019**, *12*, 3648. [CrossRef]

17. Nguyen, V.H.; Vu, C.M.; Choi, H.J.; Kien, B.X. Nanosilica extracted from hexafluorosilicic acid of waste fertilizer as reinforcement material for natural rubber: Preparation and mechanical characteristics. *Materials* **2019**, *12*, 2707. [CrossRef]

18. Murcia-Salvador, A.; Pellicer, J.A.; Rodríguez-López, M.I.; Gómez-López, V.M.; Núñez-Delicado, E.; Gabaldón, J.A. Egg by-products as a tool to remove direct Blue 78 dye from wastewater: Kinetic, equilibrium modeling, thermodynamics and desorption properties. *Materials* **2020**, *13*, 1262. [CrossRef]

19. Rizzi, V.; Gubitosa, J.; Fini, P.; Romita, R.; Nuzzo, S.; Cosma, P. Chitosan biopolymer from crab shell as recyclable film to remove/recover in batch ketoprofen from water: Understanding the factors affecting the adsorption process. *Materials* **2019**, *12*, 3810. [CrossRef]

20. Nguyen, D.D.; Vu, C.M.; Vu, H.T.; Choi, H.J. Micron-size white bamboo fibril-based silane cellulose aerogel: Fabrication and oil absorbent characteristics. *Materials* **2019**, *12*, 1407. [CrossRef]

21. Xue, J.; Zhang, K.; He, Z.; Zhao, W.; Li, W.; Xie, D.; Luo, B.; Xu, K.; Zhang, H. Rapid immobilization of simulated radioactive soil waste using self-propagating synthesized $Gd_2Ti_2O_7$ pyrochlore matrix. *Materials* **2019**, *12*, 1163. [CrossRef] [PubMed]

22. Medyńska-Juraszek, A.; Ćwieląg-Piasecka, I.; Jerzykiewicz, M.; Trynda, J. Wheat straw biochar as a specific sorbent of cobalt in soil. *Materials* **2020**, *13*, 2462. [CrossRef] [PubMed]

23. Petrella, A.; Petrella, M.; Boghetich, G.; Petruzzelli, D.; Ayr, U.; Stefanizzi, P.; Calabrese, D.; Pace, L. Thermo-acoustic properties of cement-waste-glass mortars. *Proc. Inst. Civ. Eng. Constr. Mater.* **2009**, *162*, 67–72. [CrossRef]

Article

Wheat Straw Biochar as a Specific Sorbent of Cobalt in Soil

Agnieszka Medyńska-Juraszek [1,*], Irmina Ćwieląg-Piasecka [1], Maria Jerzykiewicz [2] and Justyna Trynda [3]

[1] Institute of Soil Science and Environmental Protection, Wroclaw University of Environmental and Life Sciences, Grunwaldzka 53, 50-357 Wrocław, Poland; irmina.cwielag-piasecka@upwr.edu.pl

[2] Faculty of Chemistry, Wroclaw University, Joliot-Curie 14, 50-383 Wrocław, Poland; maria.jerzykiewicz@chem.uni.wroc.pl

[3] Department of Experimental Biology, Wroclaw University of Environmental and Life Sciences, Norwida 27b, 50-375 Wrocław, Poland; justyna.trynda@upwr.edu.pl

* Correspondence: agnieszka.medynska-juraszek@upwr.edu.pl

Received: 28 April 2020; Accepted: 24 May 2020; Published: 28 May 2020

Abstract: There is an urgent need to search for new sorbents of pollutants presently delivered to the environment. Recently biochar has received much attention as a low-cost, highly effective heavy metal adsorbent. Biochar has been identified as an efficient material for cobalt (Co) immobilization from waters; however, little is known about the role of Co immobilization in soil. Hence, in this study, a batch experiment and a long-term incubation experiment with biochar application to multi-contaminated soil with distinct properties (sand, loam) were conducted to provide a brief explanation of the potential mechanisms of Co (II) sorption on wheat straw biochar and to describe additional processes that modify material efficiency for metal sorption in soil. The soil treatments with 5% (v/w) wheat straw biochar proved to be efficient in reducing Co mobility and bioavailability. The mechanism of these processes could be related to direct and indirect effects of biochar incorporation into soil. The FT-IR analysis confirmed that hydroxyl and carboxyl groups present on the biochar surface played a dominant role in Co (II) surface complexation. The combined effect of pH, metal complexation capacity, and the presence of Fe and Mn oxides added to wheat straw biochar resulted in an effective reduction of soluble Co (II), showing high efficiency of this material for cobalt sorption in contaminated soils.

Keywords: biochar; wheat straw; sorbent; cobalt; copper; soil

1. Introduction

In recent decades, industry's reliance on cobalt as a material essential for enabling technological development has caused considerable growth in the use of cobalt and accidental release of this metal into the environment. Metal ore mining and the smelting process (mainly copper, zinc, lead, and cobalt), alloys and chemicals containing cobalt (Co), sewage effluents, and urban and agricultural runoff (phosphate fertilizers and pesticides) [1] have been described as the main sources of cobalt pollution in the environment. Since 50% of cobalt produced globally is found in rechargeable lithium-ion batteries [2], the electronic devices industry and its reliance on cobalt should be considered to be a new environmental threat. The scale of cobalt release to environmental components is not well recognized. Elevated concentrations of cobalt in soil and groundwater occur locally depending on the local geology or atmospheric deposition from metal ore mining and smelting sites, making the problem insignificant. However, cobalt can be easily transferred by air deposition into soil or leached to groundwater, affecting crop quality and food safety [3]. Cobalt plays a significant role as a constituent of vitamin B12, however, excessive exposure has been shown to induce various adverse health effects [4]. The occurrence of xenobiotic with unknown impacts on the environment and human health brings new challenges to risk reduction of elemental transfer to food chains. Among the variety of methods of soil remediation, the application of chemical amendments to polluted soil, leads to reduced environmental risks of heavy metals through several chemical mechanisms including adsorption [5], precipitation [6], and complexation [7]. According to previous studies, inorganic materials such as lime, zeolite, and phosphate are effective for heavy metals immobilization [8]. There are also organic amendments which can achieve similar efficacy, such as peat, brown coal, and biosolid compost [9–12]. The application of organic materials is a main strategy for remediation of soil polluted with heavy metals, but this procedure should be considered carefully in the case of Co polluted soils, as raw organic materials can increase mobility of this element in soil due to formation of organic chelates [9]. In a search for the most desired and efficient remediation material for cobalt, biochar should be considered. Due to the presence of a highly-porous structure [13], various functional groups (e.g., carboxyl, hydroxyl, and phenolic groups) [14] on biochar show a great affinity for metal cations [15]. The composition and biochar stability establish the sorption properties of the material. Surface functional groups present on organic carbonaceous phases play the most important role, as they decide about properties of the biochar that are important for heavy metal sorption such as pH, negative charge on the surface, the cation exchange, and surface complexation potential for metals [16]. In addition to organic components, biochars also contain mineral components such as quartz, calcite, sylvite, periclase, and whitlockite [17]. The mineral components of biochars can work as additional sorption sites for metals, ion exchange [18,19], surface complexation [20] and formation of metal precipitates [21] by releasing soluble ions, which include phosphates, sulphates, and carbonates [22,23]. A comparison with other forms of carbonaceous sorbents shows that biochar is a promising adsorbent with lower cost for metal removal from water. Much research has recently been conducted to explore biochar efficiency for heavy metal, including Co removal from an aqueous solution [24–27]. Most of this research has provided sorption mechanisms for metals as a group, however, a comparison of mechanisms for removal of different metals is necessary to describe biochar capacity for heavy metal sorption. As different metals are present in the environment in different species or valence states under different pH or red-ox conditions, the main mechanisms for their sorption could be different [28]. Cobalt most commonly occurs in the soil as Co (II) and Co (III) ions, however bioavailability and the potential environmental risk of this species in soil is distinct. The behavior of Co, in soils, is influenced to a large degree by the presence of Mn and Fe oxides which are known to have a great affinity for Co, as most of the Co (up to 79%) has been found strongly associated with Fe and Mn oxyhydroxides in soils [9,29]. Co (II) is highly soluble in water, potentially very mobile [30], and bioavailable [4]. Co (III) occurs mainly through surface oxidation of Co (II) on oxyhydroxide minerals [29], which is an important process reducing Co mobility and bioavailability in soils. Many different methods have been dedicated to estimate the efficiency of the material for metal sorption, however, sorbent efficiency could be different in soil as compared with aqueous solutions, as soil properties such texture, organic matter content, pH, or redox conditions have an influence on metal mobility and bioavailability of metal ion, making this matrix more dynamic and interactive. Described soil properties can be modify by sorbent when added to soil. As well as the biochar properties can be changed over time by weathering, leaching, oxidation, or biodegradation processes after remaining in the soil for a period of time. This makes material evaluation for remediation purposes more complex. The present study focuses on wheat straw biochar efficiency for Co sorption in soil. The batch experiment and the long-term incubation experiment with biochar application to contaminated soil provide a brief explanation of the potential mechanisms of Co sorption on wheat straw biochar and describe additional processes that modify material efficiency for metal sorption in soil.

2. Materials and Methods

2.1. Biochar Characteristics

Biochar was produced from wheat straw (WSBC) at the pyrolysis temperature of 550 °C and time remaining in the reactor 60 s. The BET surface area, cation exchange capacity (CEC), pH in deionized water, CNHSO elemental composition, ash and carbonates content (% volume/dry weight), exchangeable cations and anions content (Ca^{2+}, Mg^{2+}, K^+, Na^+, P, NH_4^+, NO_3^-), and the total contents of trace elements (Co, Mn and Fe) were determined to describe the properties of the material. The total surface area was determined using a BET (Brunauer, Emmett and Teller) specific surface area analyzer Gemini VII 2390 Series (Micrometrics Instruments Corporation, Norcross, GA, USA). The cation exchange capacity and exchangeable cations (Ca^{2+}, Mg^{2+}, K^+, Na^+, and NH_4^+) were determined according to the modified method described by Munera-Echeverri et al. [31] and analyzed on a microwave plasma-atomic emission spectrometer MP-AES 4200 (Agilent Technologies, Santa Clara, CA, USA). Exchangeable P was analyzed on the MP-AES 4200 after sample extraction according to Olsen et al. [32,33], as 0.5 M sodium bicarbonate ($NaHCO_3$) solution at a pH of 8.5 had a similar pH to WSBC. This extractant decreased calcium in solution (through precipitation of calcium carbonate), and this decrease enhanced the dissolution of Ca-phosphates. The nitrate content was analyzed according to the ISO 14256-1:2003 procedure on a UV-Vis Cary 60 (Agilent Technologies, Santa Clara, CA, USA). The pH values were measured at a ratio of 1:5 (w/v) in deionized water after the sample was shaken for 1 h at 130 rpm with a calibration check pH meter (Mettler Toledo, Columbus, OH, USA). The ash content was determined by weight loss after combustion at 750 °C for 6 h in a muffle furnace according to ASTM D7348-13 [34]. $CaCO_3$ was determined following the Scheibler method with a calcimeter [35]. The elemental composition (CHNSO) was analyzed on a CHNS analyzer (CE Instruments, Hindley Green, UK), and the O content was calculated from the difference. The total content and exchangeable forms of metals (Co, Cu, Fe, and Mn) were analyzed on a microwave plasma-atomic emission spectrometer MP-AES 4200 (Agilent Technologies, Santa Clara, CA, USA), respectively, after microwave sample digestion in 70% nitric acid (1:10 w/v ratio) in a digestion microwave system StartD (Milestone Srl.Sorisole, Italy) and extraction with deionized water (1:25 w/v ratio).

2.2. Metal Sorption Mechanism Analysis

Previous studies of Co sorption in soil have showed that Co and Cu as divalent cations can compete for sorption sites, especially in excess of Cu^{2+} in multi-contaminated soils, as described by Muyumba et al. [36]. In the metal sorption experiment, both metals were used to simulate natural conditions in contaminated soils and possible interactions of the Co and Cu ions. The FTIR spectroscopy was used to compare potential changes in the functional groups of metal-loaded biochars with the biochar samples before the Co or Cu sorption. The Co and Cu sorption on the wheat straw biochar was determined by the simplified batch equilibrium method according to the OECD 2000/106 protocol [37]. To compare the effect and probable interaction between cobalt and copper that could occur in multi-contaminated soils the following three solution were used in the batch experiment: (1) Co (II) acetate, (2) Cu (II) acetate, and (3) Co (II) + Cu (II) mix of both salts. Briefly, 5 g of each salt was diluted in 500 mL of deionized water and set overnight to reach equilibrium. The pH of each solution was measured before and after the batch experiment to determine if the pH change occurred after biochar BC addition which would affect sorption conditions. One gram of wheat straw biochar and 20 mL of each solution were placed in 50 mL volume polypropylene falcon centrifuge tubes. All samples were prepared in three replicates of each treatment. The sealed samples were placed on the rotary shaker Multi RS-60 (Biosan, Riga, Latvia) at 80 rpm and 22 ± 0.5 °C for 24 h. Sorption equilibrium was reached within less than 24 h. Then, samples were centrifuged for 25 min at 10,000 rpm to separate the biochar from the solution according to procedure described by Ćwielag-Piasecka et al. [38]. The biochar samples were washed three times with 20 mL of deionized water and prefiltrated on Munktell No.

2 filter papers (Ahlstrom Munksjö, Helsinki, Finland) to rinse off excess metal cations. The biochar samples were dried in an oven drier at 60 °C for 6 h to prepare pellets for Fourier transform infrared spectra (FT-IR). The FT-IR analysis of the wheat straw biochar samples were recorded using a Vertex 70 FT-IR spectrometer (Bruker, Billerica, MA, USA) on KBr pellets (about 1 mg sample for 400 mg of KBr) according to the standard method used for sample preparation for FT-IR spectra analysis. The incubation experiment with multi-contaminated soils was a pot experiment with 24, four-liter pots (approximately 3 kg of soil each). Two soil types, sand and loam, were set as 12 control pots, six for each soil type. A similar 12 pots were amended with a dose of BC (5.0% w/v), six for each soil type. The soil mixtures were incubated for two years, keeping the humidity of the pots at 60% of maximum water holding capacity. After the time period, the soil samples were collected from each pot, air dried, and sieved (<2 mm), and sequential extraction of Co was performed. The existence of possible precipitates of Co after metal sorption was checked using a scanning electron microscope (SEM) (Bruker, Billerica, MA, USA) coupled with an energy dispersive X-ray analyzer (EDX) (Bruker, Billerica, MA, USA). The biochar particles for SEM-EDX (scanning electron microscopy with energy dispersive X-ray spectroscopy) analysis were separated from the soil by progressive sieving of the soil-biochar mixture. Biochar fraction <1 mm was used to determine elemental composition of the material surface and the distribution of ions [39]. For the morphological observations, SEM was applied according to Michalak et al. [40].

2.3. Soils Sampling

Ten soil samples were collected from the topsoil horizon (0–20 cm) at afforested sites at different distances and locations from the copper smelter in SW Poland (16°01′40″N, 51°45′09″E), expecting cobalt enrichment from smelter emissions. In the samples, pre-scanning studies for a wide range of heavy metals, including cobalt, were done on a microwave plasma-atomic emission spectrometer MP-AES 4200 (Agilent Technologies, Santa Clara, CA, USA) after microwave sample digestion in 70% nitric acid (1:10 w/v ratio). The total concentration of cobalt remained in a wide range from 4.5 to 74 mg/kg, depending on the distance from the smelter and the soil type. Two soils with the highest Co concentrations, differing in texture, were chosen for the incubation experiment. Large soil samples, 15 kg each, were collected of Cutanic Luvisol and Fulvic Brunic Arenosol according to the FAO (Food and Agriculture Organization of the United Nations) guidelines [41] at a distance of approximately 2 km from the potential emission source. The copper smelter has been in operation since 1968, becoming one of the most important sources of airborne heavy metal pollution in this area. Polymetallic deposits contain about 1.4% of Cu, however significant amounts of Co-barring minerals can be also found in the ore, mainly Cu-Co, Ni-Co, and As-Co minerals. As the ore is excavated, floated, and smelted, these processes become significant sources of airborne cobalt emission. Unfortunately, no data on cobalt and other rare metals are available for this area. The basic soil properties were examined using common methods described by [42]. To simplify the descriptions further, soil samples were named L for Cutanic Luvisol and S for Fulvic Brunic Arenosol, as well as C for the control, and WSBC for wheat straw biochar. The initial physicochemical properties and total content of cobalt of the studied soils are presented in Table 1. The Cutanic Luviosol Control (LC) soil sample was loamy and neutral in nature, non-saline, with average European soil total carbon content and high CEC. The Fulvic Brunic Arenosol Control (SC) soil sample was sandy and acidic, non-saline, with typical sandy soil low carbon content and low CEC. The total content of Co in the LC soil sample was higher than that in the SC soil sample, respectively, 67 and 26 mg/kg. In both soil samples, the total Co concentrations were higher than that for European soils, i.e., <9.3 mg/kg [8,38,39]. Naturally, higher Co contents were observed in soils around ore deposits, as the distance from the smelter to mining sites is about 40 km, we indicated this element as an airborne pollution from the copper smelter.

Table 1. Selected properties of soils from the incubation experiment.

Sample	Depth	Textural Group	Clay %	C_{org} %	pH_{H2O}	CEC $cmol_c$/kg	Co_{tot} mg/kg
LC	0–20 cm	Silt loam	4	$1.12^{1} \pm 3.5^{2}{}_{a}$	$6.9 \pm 11.0_a$	$58 \pm 8.3_a$	$67.0 \pm 2.3_a$
SC	0–20 cm	Loamy sand	<1	$0.98 \pm 4.6_a$	$3.9 \pm 20.1_a$	$5.5 \pm 14_a$	$26.0 \pm 1.8_a$
L + 5% WSBC	0–20 cm	-	-	$1.36 \pm 3.8_b$	$7.3 \pm 14.6_a$	$63 \pm 9.2_a$	$67.2 \pm 1.7_a$
S + 5% WSBC	0–20 cm	-	-	$1.17 \pm 2.7_b$	$4.6 \pm 12.5_b$	$6.3 \pm 5.8_a$	$26.2 \pm 1.9_a$

[1] Mean values (n = 6) and [2] RSD values in % (n = 6). Different lowercase letters (a and b) indicate significant differences between WSBC treated and untreated soil within each soil type ($p < 0.05$).

2.4. Cobalt Immobilization Analysis in Soil

After soils were incubation with 5% (w/v) wheat straw biochar, the (Community Bureau of Reference (BCR) sequential extraction procedure was applied to measure the following four fractions of cobalt in the tested soil: exchangeable and bound to carbonates (Fraction 1), reducible or bound to Fe and Mn-oxides (Fraction 2), oxidizable or bound to organic substances (Fraction 3), and residual (Fraction 4). Acetic acid, hydroxyl ammonium chloride, hydrogen peroxide plus ammonium acetate, and aqua regia stages of the sequential extraction procedure were applied to the soil samples, respectively [43–45]. The contents of cobalt, extracted during the BCR procedure, were measured on a microwave plasma-atomic emission spectrometer MP-AES 4200 (Agilent Technologies, Santa Clara, CA, USA). Data are provided as an average result from triplicate with the relative standard deviation (RSD), calculated by MP Expert Software Agilent Technologies. The maximum relative standard deviation (RSD) between replicates was set to 5%. Values that were above 5% were not included in the statistical analyses. To avoid analytical errors, standard solutions (from LGC Standards Ltd., UK) for MP-AES 4200 were used for calibration and certified reference materials as follows: RTH 953 Heavy Clay Soil from LGC Promochem (LGC Standards Ltd., Teddington, UK), total Co content 14.7 mg/kg and CRM055 (Honeywell Fluka, Charlotte, NC, USA) with Co content of 97 mg/kg were analyzed with every sample set. The recovery of Co from Certified Reference Material (CRM) was 89–94% and the maximum values of RSD were 2.6%. Detection limits were 0.01 mg/kg of Co in the soil samples.

2.5. Statistical Analysis

The metal sorption batch experiments were performed in triplicate and the soil sorption experiments were performed in six replicates. The data are presented as the mean values with the relative standard deviation (RSD). Student's *t*-tests were used to test for significant differences in cobalt fractions between biochar treated and untreated soils ($p < 0.05$). The obtained data were compiled using Microsoft Excel 2016 and Statistica Statsoft 13.3.

3. Results

3.1. Sorbent Characteristics

The elemental composition of the tested wheat straw biochar is presented in Table 2. The results indicated that the C content in the WSBC was 63.3%, followed by O content of 33.46%, N content of 0.74%, H content of 2.2%, and very low S content of 0.037%. The calculated H/C and O/C molar ratios are the indicators of BC aromaticity and polarity, respectively. It is assumed that BCs produced at a temperature higher than 400 °C should be characterized by an H/C ratio lower than 0.5 and decrease with the raising pyrolysis temperature below 0.3, which is an indicator of highly aromatic ring systems. In the case of the investigated BC, produced at 550 °C, the H/C ratio falls within the 0.3–0.5 range which could indicate a decreased fraction of original wheat residues. The obtained BC molar ratios (Table 2) emphasize the presence of aromatic structural features and reduced content of O-containing polar functional groups on BC surface (low molar O/C ratio and polarity index) [38].

Table 2. Elemental composition of pure and spiked with metal salts wheat straw biochar.

Sample	C	N	H	S	O	H/C	O/C	C/N
	% (w/dw)					molar		
WSBC	$63.6^{1} \pm 1.6^{2}$	0.74 ± 3.5	2.2 ± 5.2	0.037 ± 8.8	33.46 ± 0.8	0.34	0.52	85.9
WSBC + Cu (II)	63.8 ± 3.4	0.62 ± 3.2	1.9 ± 5.2	0.041 ± 7.4	33.05 ± 1.9	0.29	0.51	102.9
WSBC + Co (II)	64.5 ± 3.4	0.64 ± 3.2	2.13 ± 1.8	0.041 ± 6.8	33.45 ± 1.2	0.33	0.51	100.7
WSBC + Cu (II) + Co (II)	63.9 ± 0.7	0.65 ± 3.7	2.02 ± 5.8	0.042 ± 6.9	$33.32 + 1.4$	0.31	0.52	98.3

[1] mean values (n = 3) and [2] RSD values (n = 3).

Sorption properties of tested wheat straw biochar are given in Table 3. The average specific surface area (SSA) of the tested WSBC was 256 g/m^2, with the cation exchange capacity (CEC) of 63 cmol$_c$/kg. The total ash content was 32.4% and the contribution of calcium carbonates was 3.07% (w/dw), however almost 90% of exchangeable cations in CEC was a K$^+$, followed by Ca^{2+}, Mg^{2+}, and a very low content of Na$^+$ and NH$_4{}^+$ (Table 3). The content of exchangeable phosphorus, mainly in forms represented by Ca-phosphates (Olsen P) was 265 mg/kg, which was less than 9% of the total phosphorus in tested wheat straw biochar. The content of NH$_4{}^+$ and NO$_3{}^-$ was very low, and less than 0.1% of total nitrogen was in exchangeable forms after sample extraction with 1 M KCl. The total content of sulfur was 0.037–0.042% (Table 1), suggesting that the contribution of sulfate forms in WSBC was negligibly low. The low content of cation and anion in exchangeable forms after extraction with weak extractants could be attributed to the very high pH (9.86) of biochar. Similar to multi-contaminated matrixes such as the soils from the copper smelter area, competition, between Co (II) and other metal cations, can occur and some of the metals present on the biochar surface can be exchanged, and therefore the potential contribution in this process of Cu, Fe, and Mn was analyzed. The results showed that only 2.6% of Cu, 2.8% of Fe, and 4.1% of Mn in the tested biochar were in readily exchangeable forms. The total content of Co and exchangeable Co forms showed very low potential contributions of biochar-derived Co, in the sorption/desorption processes (Table 4). However, high content of exchangeable Cu^{2+} in the tested soil could induce competition between both divalent cations for sorption sites on the biochar surface, which was shown in the batch experiment.

Table 3. Properties of wheat straw biochar.

pH$_{H2O}$	SSA	Ash$_{tot}$	CaCO$_3$	CEC	Ca^{2+}	Mg^{2+}	K$^+$	Na$^+$	NH^{4+}	NO$_3{}^-$	P$_{ex}$ *
	m^2/g	%	%		cmol$_c$/kg					mg/L	
9.86 [1]	256 ±	32.4 ±	3.07 ±	63 ±	8.38 ±	5.22 ±	58.1 ±	0.76 ±	0.025	4 ±	265 ±
± 6.9 [2]	3.3	13.3	19.1	22.4	14.1	19.6	26.7	25.3	± 15.1	25.1	30.1

[1] Mean value (n = 6) and [2] RSD value (n = 6), * Olsen P.

Table 4. Total and exchangeable forms of metals in wheat straw biochar.

Co$_{tot}$	Cu$_{tot}$	Fe$_{tot}$	Mn$_{tot}$	Co$_{ex}$	Cu$_{ex}$	Fe$_{ex}$	Mn$_{ex}$
			mg/kg				
$1.1^{1} \pm 3.2^{2}$	15.3 ± 4.8	1156 ± 1.2	260 ± 3.5	0.03 ± 4.3	0.4 ± 2.7	32 ± 1.8	10.8 ± 3.7

[1] Mean value (n = 6) and [2] RSD value (n = 6).

3.2. Metal Ions Sorption on Biochar

The FT-IR analysis showed differences between pure and spiked with Co (II), Cu (II), and mix Co (II) + Cu (II) biochars (Figure 1). From the presented spectra, the most probable mechanism of metal ions binding can be related to the oxygen containing groups on wheat straw biochar surface, as the most characteristic changes occurred at vibrations 3428, 1624, and 1420 cm^{-1}. The metal ions in metal spiked biochars decreased the intensity of the peaks at 3428 cm^{-1} stretching vibrations of the OH (H-bonding) groups. This change confirms that the O-H groups take part in Co (II) and Cu (II) complexation on wheat straw biochar surface. The carboxyl peak observed for a pure wheat straw biochar at 1624 cm^{-1} was shifted to much smaller values, i.e., 1583 cm^{-1} or 1570 cm^{-1} in the spectra of BC treated with salts (Figure 1).

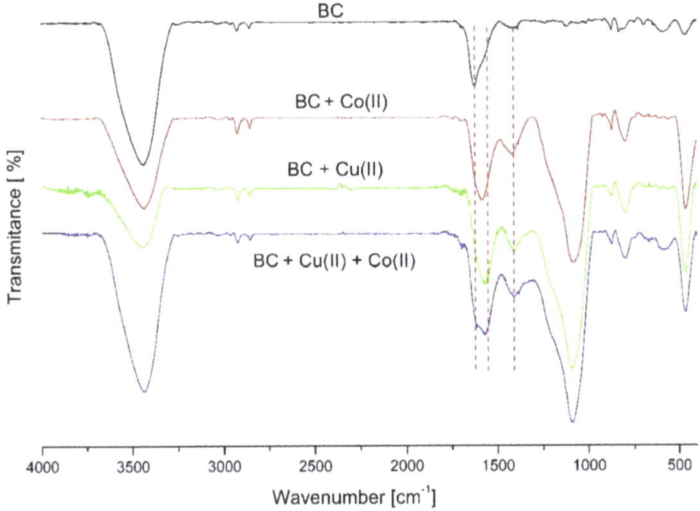

Figure 1. FT-IR spectra of investigated pure wheat straw biochar, biochar spiked with Co (II) salt, biochar spiked with Cu (II) salt, and biochar spiked with a mix of Co (II) and Cu (II) salts.

The decrease in the wavenumber of the peak 1624 cm^{-1}, characteristic for C=O carboxylic group, can be explained by the interaction with Co (II) and Cu (II) ions with free carboxyl groups on the biochar surface and change to carboxylates, which indicates the important role of carboxyl groups on the biochar surface in metal binding. To provide more details about the type of the metal binding to carboxylic group on the biochar surface at 1624 and 1420 cm^{-1}, a calculation of Δ according to Nakamoto [46] was performed (Table 5).

Table 5. Calculated values of Δυ (COO$^-$).

Sample	ν_a(COO$^-$) (cm^{-1})	ν_s(COO$^-$) (cm^{-1})	$\Delta = \nu_a$(COO$^-$) $- \nu_s$(COO$^-$)
WSBC	1624	1420	204
WSBC + Cu (II)	1583	1419	164
WSBC + Co (II)	1570	1409	161
WSBC + Cu (II) + Co (II)	1574	1411	163

According to the calculations, carboxylate could coordinate metal ions on three different modes, i.e., unidentate, chelating (bidentate), and bridging [47]. The calculated values of Δ indicate creation of bridging complexes, where two metal ions are involved in the binding of one carboxylic group. Some of the metal coordination sites can be associated with H$_2$O or OH groups. The results of the FT-IR analysis also showed that metal ions complexation could be related to an abundance of carbonates and polysaccharides such as moieties in biochar, since changes of peak at 1080 cm^{-1} related to Si–O, C–O, and S=O groups were observed in metal spiked biochar. Very low concentrations of sulfur in biochar (Table 2) can limit metal ion coordination with S=O groups on the BC surface, however, the Si–O groups can be involved in the process. The very strong peak, 464 cm^{-1}, can be attributed to vibrations of many moieties. This band could appear when H$_2$O is one of the ligand in the complex or Cu–O–H deformations [46].

3.3. Cobalt Immobilization in Soils

Wheat straw biochar was applied to soils to verify the hypothesis that application of this material can immobilized cobalt ions in soil. Copper fractionation for tested soils was described in the previous paper [48] and is not be discussed in these results. Figure 2 shows the contribution of Co forms in the control sand and the loam and soils treated with 5% (v/dw) wheat straw biochar. Regardless of the type of soil, the Co contribution of the individual fractions was similar as follows: F1 < F2 < F3 < F4; however, significant differences ($p < 0.05$) were observed in fraction content between soils. In both tested soils, mobility of Co was high (more than 10%) as compared with other analyzed metals in tested soils [48].

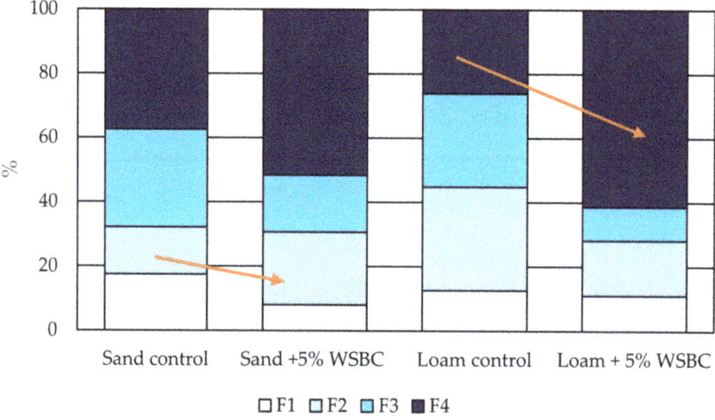

Figure 2. Cobalt speciation in treated and untreated soils. Cobalt fractions: F1, exchangeable and bound to carbonates; F2, reducible or bound to Fe and Mn-oxides; F3, oxidiziable or bound to organic substances; and F4, residual.

The application of 5% (w/v) wheat straw biochar affected cobalt speciation in both tested soils. In sand + 5% WSBC treatment, a significant decrease in exchangeable cobalt Fraction F1 (F1) from 17.4% to 7.3% was determined. The observed reduction in Co content in F1 was mainly balanced by their increased content in Fraction 2 (F2), reducible or bound to Fe and Mn-oxides (Figure 2) by 9.4%. Some of the cobalt forms were also shifted from Fraction 3 (F3), oxidiziable or bound to organic substances and Fraction 4 (F4), becoming residual fraction non-bioavailable or not prone to leaching [49]. In the loam + 5% WSBC treatment, the biochar application did not decrease Co speciation in Fraction 1, however significant amounts of Co were shifted from Fractions 2 and 3 to Fraction 4, increasing cobalt in residual fraction by 40%.

3.4. SEM-EDX Analysis of Biochar

The application of SEM-EDX proved that the metal ions can be bound on the biochar surface. Examination of the SEM-EDX element map, Figure 3, presents the relative increase in surface concentration of Co on biochar particle (<1 mm), shown by the image brightness after biochar incubation in contaminated soil. The location of Co "hot spots" on the biochar surface corresponded to sites with increased surface concentrations of Fe and Mn and oxygen functional groups.

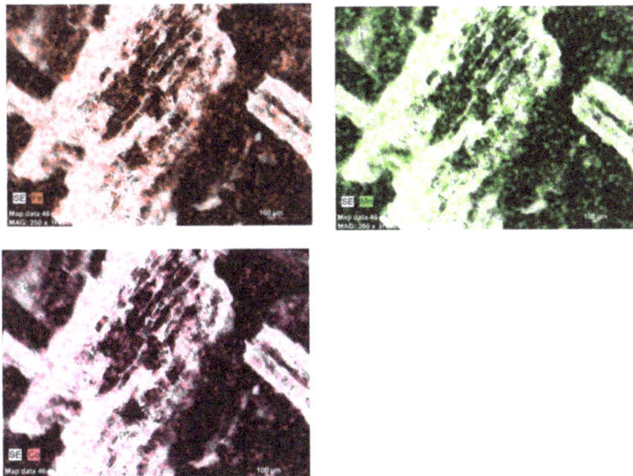

Figure 3. Scanning electron microscope (SEM) images of wheat straw biochar surface and energy dispersive X-ray spectroscopy (EDX) mapping. (**a**) Fe; (**b**) Mn; and (**c**) Co ions, distribution on the biochar produced from wheat straw (WSBC) surface.

4. Discussion

Biochar application can have a direct or indirect impact on cobalt immobilization/mobilization process in soil. Indirectly, biochar can affect soil sorption properties and pH, reducing the presence of metal in exchangeable and soluble forms in soil solution. The direct effect can be related to biochar properties such as sorption capacity, oxygen functional groups, and mineral components content (carbonates, phosphates, Mn and Fe oxides) increasing or supporting soil sorption capacity for cobalt. All mentioned processes were considered in our study. The tested wheat straw biochar had high SSA as compared with other straw-derived biochars produced under similar conditions, as porosity of biochar significantly increases between 400–600 °C [50]. Gul et al. [51] characterized wheat straw biochar with SSA from 178 to 184 m^2/g, depending on pyrolysis (slow or fast), although, in most research, lower values can be found with good efficiencies for heavy metal removal [52,53]. Cation exchange capacity was low, as this property usually decreases with the higher temperatures of pyrolysis [1]. Wheat straw biochar had a very high pH, most likely due to the much higher potassium content found in the straw biochars as compared with wood derived materials, which was also described in another study [54]. The wheat straw biochar had a high ash and carbonates content, although very low content of nitrates, sulphates, and phosphates, which has also been indicated by other authors studying straw-derived biochars [51,54]. Biochar properties affect soil properties when applied to soil [50], modifying conditions of heavy metal mobility (mainly CEC and pH). Wheat straw biochar failed to change the soil CEC significantly ($p > 0.05$) in both tested soils. Normally, biochars are considered to develop more oxygen-containing functional groups, and hence increase CEC and negative charge of soils [13,19,55]. Our study suggested that an increase of soil functional groups after WSBC was not enough to change sorption properties of soil, which was in agreement with a study by Qi et al. [56]. Wheat straw biochar changes soil pH significantly ($p > 0.05$), but only in acidic sandy soil, which could affect cobalt mobility indirectly. Soil pH is one of the most important environmental factors affecting sorption of toxic metal [57]. In other experiments with wheat straw biochar, cobalt sorption from aqueous solution was described as pH dependent, i.e., very low at pH values from 2.0 to 4.0 and high between pH values of 5.5 and 8.0 [43,58], similar to the pH of tested soil samples after WSBC application. A lower pH, such as in sandy soil, enhanced cobalt mobility and potential bioavailability for plants, which was also observed in other studies on acidic soils [59,60]. As our study did not show a significant impact on soil sorption capacity, we focused on biochar characterization as some direct mechanisms such as surface complexation or supported sorption by addition mineral compounds like Fe and Mn oxides, could be related to decreased mobility of cobalt in the tested soils. The results of the FTIR analysis showed that the tested wheat biochar had the capacity for Co complexation with oxygen-containing surface functional groups, mainly carboxylic C=O and hydroxyl groups H–O, similar to other divalent cations. Depending on the pH conditions, cobalt typically to other divalent cations, can be hydrated in a soil solution sharing similar sorption mechanisms, i.e., cation exchange, surface complexation, and precipitation [21]. Similar results for active Co chemisorption by hydroxyl functional groups were described by Liu et al. [61]. Sun et al. [27] showed that oxygen-containing surface functional groups, for example, C–O, C–O–C, and C=O increased biochar capability for element immobilization [62], suggesting that these groups play important roles in metal sorption. The quantities and quality of functional groups on biochar surface vary, depending on biochar production conditions and feedstock types used. However, biochars produced at higher temperatures (>500 °C) have higher surface area and porosity, but lower abundance of functional groups, primarily due to the higher degree of carbonization [28]. Zhang et al. [63] observed decreased contribution of O-alkyl carbon from 20% to 54% to 7% to 13% for wheat straw biochars, as temperature increased from 200 to 600 °C, and at >300 °C aromatic structures are dominant [64]. These findings are in agreement with the CHNSO analysis and molar ratios obtained during raw material analysis in this study, emphasizing the presence of aromatic structural features and reduced content of O-containing polar functional groups on WSBC surface.

As oxygen-containing functional groups are predominant mechanisms of divalent cations the result of our study suggests that biochars produced at lower temperatures or oxidized during pretreatments [65] have better efficiency for Co^{2+} removal as compared with materials produced at high temperatures with a more aromatic structure. However, biochar applied to soil vs. solution undergoes many abiotic and biotic processes causing sorbent oxidation, called biochar aging [66], which can result in increased sorption capacity for cations after time remaining in soil. Aging in soil leads to surface changes [67] and generation of new functional groups on its surface [68]. Uchimiya et al. [67] described that in the presence of soil, the importance of oxygen-containing groups on biochar surfaces in cation sorption strongly depended upon the inherent sorption capacity of soil. Wang et al. [66] observed that decarboxylation of surface functional groups on biochar surface, when added to soil, increased soil pH, but these groups could also affect complex metal cations in the soil solution and reduce bioavilalbility, which is in agreement with our study.

Findings of our study showed that wheat straw biochar can support soil sorption complex acting as a source of several mineral components, for example, Fe and Mn oxides [69,70], silica, carbonates, and phosphates [71], increasing cobalt immobilization/precipitation in soil. The Fe and Mn oxides were not specifically analyzed by XRF in the present study, however, interaction between the Fe and Mn ions and Co was observed after biochar incubation in soils by SEM-EDS analysis and sequential extraction of Co from tested soils where cobalt fraction bound to Fe and Mn oxides increased significantly ($p > 0.05$) in biochar treated soils. The Mn and Fe oxides and organometallic moieties such as Fe–O–C can be formed on the biochar surface during pyrolysis. Most of the Fe in biochar is present in crystalline phases ranging from zerovalent iron to ferric oxides [72]. The role of hydrous oxides of iron, manganese, and clay minerals as Co sorbent has been described recently [73–77]. As these reactions are strongly pH dependent, under alkaline conditions in loam +5% WSBC treatment, cobalt could coprecipitate as Fe and Mn secondary oxides, shifting Co to residual forms, which has also been observed in other studies [43]. Kabata-Pendias [9] described that cobalt precipitation on Mn oxide surface increased under alkaline conditions forming very stable hydroxyl species $Co(OH)_2$ which could explain the high stability of Co forms after WSBC application to loam soil. Pan et al. [24] described that Co immobilization on biochar could be dependent on mineral composition and the content of carbonates, phosphates, and calcium hydroxyapatite (CaHA), and suggested that cobalt could be rapidly exchanged with calcium decreasing element in the solution. This mechanism was also visible in the FT-IR analysis, however, analysis exchangeable cation extraction from tested soil did not indicated any significant changes. Figure 4 presents the described conceptual model and possible mechanisms of cobalt sorption on wheat straw biochar.

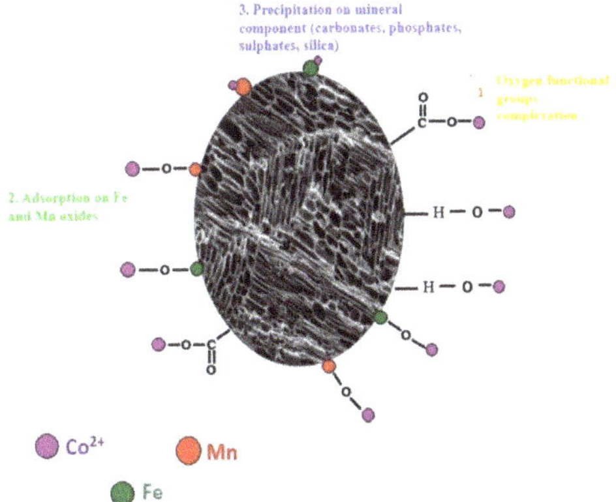

Figure 4. Conceptual model of Co adsorption mechanism on wheat straw biochar surface.

The results of our study showed that wheat straw biochar has good removal efficiencies in single-metal systems determined in a batch experiment with cobalt salts in the solution. However, the capacities of biochar for cobalt sorption can be modified in multiple-metal systems due to the competition between the heavy metals present in soil and lower stability of the material due to cation exchange and the surface oxidation process that biochar undergoes under soil conditions. As biochar can affect soil properties, changing conditions of metal immobilization, and as soil conditions can have an impact on surface properties of biochar, predictions about biochar efficiency for metal sorption in soil are difficult and need further recognition.

5. Conclusions

- The results of the experiments showed that wheat straw biochar is an efficient sorbent of Co^{2+}, decreasing mobility and availability of this element in soil.
- The dominant mechanisms of cobalt sorption on biochar surface are related to oxygen functional groups present on the biochar surface, mainly carboxylic and hydroxyl groups.
- Biochar can support soil sorption complex with mineral components such as Fe and Mn oxides which increase efficiency of cobalt immobilization.
- The efficiency of cobalt removal from soil can be modified by biochar oxidation and interactions with soil constituents which makes the process more complex as compared with cobalt removal from aqueous solutions.
- The results from this study suggest that application of biochar is a feasible strategy for remediation of cobalt contaminated soils, reducing health risks related to human exposure to Co from anthropogenic sources.
- Further and more complex studies are necessary to recognize the problems of cobalt soil pollution, potential risks, and remediation solutions, including biochar application.

Author Contributions: Conceptualization, A.M.-J.; methodology, A.M.-J and M.J.; validation, I.Ć.-P.; formal analysis A.M.-J., M.J., and J.T.; investigation, A.M.-J; writing—original draft preparation, A.M.-J.; writing—review and editing, A.M.-J., I.Ć.-P., and M.J. All authors have read and agreed to the published version of the manuscript.

Funding: This research was funded from the "Young scientists financial support programme", project no B030/0019/18, from the Wroclaw University of Environmental and Life Sciences, Poland.

Acknowledgments: Special thanks to Świdnicka Fabryka Urządzeń Przemysłowych for providing wheat straw biochar for the study.

Conflicts of Interest: The authors declare no conflict of interest.

References

1. Gál, J.; Hursthouse, A.; Tatner, P.; Stewart, F.; Welton, R. Cobalt and secondary poisoning in the terrestrial food chain: Data review and research gaps to support risk assessment. *Environ. Int.* **2008**, *34*, 821–838. [CrossRef]

2. Alves Dias, P.; Blagoeva, D.; Pavel, C.; Arvanitidis, N. *Cobalt Demand-Supply Balances in the Transition to Electric Mobility*; Publications Office of the European Union: Luxembourg, 2018; ISBN 978-92-79-94311-9.

3. Liu, B.; Huang, Q.; Su, Y.; Sun, L.; Wu, T.; Wang, G.; Kelly, R. Rice busk biochar treatment to cobalt-polluted fluvo-aquic soil: Speciation and enzyme activities. *Ecotoxicology* **2019**, *28*, 1220–1231. [CrossRef]

4. Leyssens, L.; Vinck, B.; Van Der Straeten, C.; Wuyts, F.; Maes, L. Cobalt toxicity in humans—A review of the potential sources and systemic health effects. *Toxicology* **2017**, *387*, 43–56. [CrossRef]

5. Beesley, L.; Moreno-Jiménez, E.; Gomez-Eyles, J.; Harris, E.; Robinson, B.; Sizmur, T. A review of biochars' potential role in the remediation, revegetation and restoration of contaminated soils. *Environ. Pollut.* **2011**, *159*, 3269–3282. [CrossRef]

6. Park, J.; Lamb, D.; Paneerselvam, P.; Choppala, G.; Bolan, N.; Chung, J. Role of organic amendments on enhanced bioremediation of heavy metal(loid) contaminated soils. *J. Hazard. Mater.* **2011**, *185*, 549–574. [CrossRef]

7. Cao, Y.; Gao, Y.; Qi, Y.; Li, J. Biochar-enhanced composts reduce the potential leaching of nutrients and heavy metals and suppress plant-parasitic nematodes in excessively fertilized cucumber soils. *Environ. Sci. Pollut. Res.* **2018**, *25*, 7589–7599. [CrossRef]

8. Karczewska, A.; Gałka, B.; Dradrach, A.; Lewińska, K.; Mołczan, M.; Cuske, M.; Gersztyn, L.; Litak, K. Solubility of arsenic and its uptake by ryegrass from polluted soils amended with organic matter. *J. Geochem. Explor.* **2017**, *182*, 193–200. [CrossRef]

9. Kabata-Pendias, A.; Pendias, H. *Trace Elements in Soils and Plants*, 3rd ed.; CRC Press: Boca Raton, FL, USA, 2001; ISBN 978-0-8493-1575-6.

10. Beesley, L.; Moreno-Jiménez, E.; Gomez-Eyles, J. Effects of biochar and greenwaste compost amendments on mobility, bioavailability and toxicity of inorganic and organic contaminants in a multi-element polluted soil. *Environ. Pollut.* **2010**, *158*, 2282–2287. [CrossRef]

11. Gonzaga, M.; Mackowiak, C.; Quintão de Almeida, A.; Wisniewski, A.; Figueiredo de Souza, D.; da Silva Lima, I.; Nascimento de Jesus, A. Assessing biochar applications and repeated *Brassica juncea* L. production cycles to remediate Cu contaminated soil. *Chemosphere* **2018**, *201*, 278–285. [CrossRef]

12. Karczewska, A.; Chodak, T.; Kaszubkiewicz, J. The suitability of brown coal as a sorbent for heavy metals in polluted soils. *Appl. Geochem.* **1996**, *11*, 343–346. [CrossRef]

13. Zhang, L.; Tang, S.; Jiang, C.; Jiang, X.; Guan, Y. Simultaneous and Efficient Capture of Inorganic Nitrogen and Heavy Metals by Polyporous Layered Double Hydroxide and Biochar Composite for Agricultural Nonpoint Pollution Control. *ACS Appl. Mater. Interfaces* **2018**, *10*, 43013–43030. [CrossRef] [PubMed]

14. Uchimiya, M.; Chang, S.; Klasson, K. Screening biochars for heavy metal retention in soil: Role of oxygen functional groups. *J. Hazard. Mater.* **2011**, *190*, 432–441. [CrossRef] [PubMed]

15. Dai, S.; Li, H.; Yang, Z.; Dai, M.; Dong, X.; Ge, X.; Sun, M.; Shi, L. Effects of biochar amendments on speciation and bioavailability of heavy metals in coal-mine-contaminated soil. *Hum. Ecol. Risk Assess. Int. J.* **2018**, *24*, 1887–1900. [CrossRef]

16. Mukherjee, A.; Zimmerman, A.; Harris, W. Surface chemistry variations among a series of laboratory-produced biochars. *Geoderma* **2011**, *163*, 247–255. [CrossRef]

17. Qi, F.; Yan, Y.; Lamb, D.; Naidu, R.; Bolan, N.; Liu, Y.; Ok, Y.; Donne, S.; Semple, K. Thermal stability of biochar and its effects on cadmium sorption capacity. *Bioresour. Technol.* **2017**, *246*, 48–56. [CrossRef]

18. Sizmur, T.; Quilliam, R.; Puga, A.; Moreno-Jiménez, E.; Beesley, L.; Gomez-Eyles, J. Application of Biochar for Soil Remediation. In *SSSA Special Publications*; Guo, M., He, Z., Uchimiya, S., Eds.; American Society of Agronomy and Soil Science Society of America: Madison, WI, USA, 2015; pp. 295–324, ISBN 978-0-89118-967-1.

19. Wang, B.; Ran, M.; Fang, G.; Wu, T.; Ni, Y. Biochars from Lignin-rich Residue of Furfural Manufacturing Process for Heavy Metal Ions Remediation. *Materials* **2020**, *13*, 1037. [CrossRef]

20. Qian, L.; Chen, B. Interactions of Aluminum with Biochars and Oxidized Biochars: Implications for the Biochar Aging Process. *J. Agric. Food Chem.* **2014**, *62*, 373–380. [CrossRef]

21. Li, H.-F.; Gray, C.; Mico, C.; Zhao, F.-J.; McGarth, S. Phytotoxicity and bioavilibility of cobalt for plants in a range of soil. *Chemosphere* **2009**, *75*, 979–986. [CrossRef]

22. Inyang, M.; Gao, B.; Yao, Y.; Xue, Y.; Zimmerman, A.; Pullammanappallil, P.; Cao, X. Removal of heavy metals from aqueous solution by biochars derived from anaerobically digested biomass. *Bioresour. Technol.* **2012**, *110*, 50–56. [CrossRef]

23. Xia, S.; Song, Z.; Jeyakumar, P.; Bolan, N.; Wang, H. Characteristics and applications of biochar for remediating Cr(VI)-contaminated soils and wastewater. *Environ. Geochem. Health* **2019**. [CrossRef]

24. Pan, X.; Wang, J.; Zhang, D. Sorption of cobalt to bone char: Kinetics, competitive sorption and mechanism. *Desalination* **2009**, *249*, 609–614. [CrossRef]

25. Pipíška, M.; Richveisová, B.; Frišták, V.; Horník, M.; Remenárová, L.; Stiller, R.; Soja, G. Sorption separation of cobalt and cadmium by straw-derived biochar: A radiometric study. *J. Radioanal. Nucl. Chem.* **2017**, *311*, 85–97. [CrossRef]

26. Vilvanathan, S.; Shanthakumar, S. Column adsorption studies on nickel and cobalt removal from aqueous solution using native and biochar form of *Tectona grandis*. *Environ. Prog. Sustain. Energy* **2017**, *36*, 1030–1038. [CrossRef]

27. Sun, X.; Zhong, T.; Zhang, L.; Zhang, K.; Wu, W. Reducing ammonia volatilization from paddy field with rice straw derived biochar. *Sci. Total Environ.* **2019**, *660*, 512–518. [CrossRef]

28. Li, H.; Dong, X.; da Silva, E.; de Oliveira, L.; Chen, Y.; Ma, L. Mechanisms of metal sorption by biochars: Biochar characteristics and modifications. *Chemosphere* **2017**, *178*, 466–478. [CrossRef]

29. Ma, Y.; Hooda, P. Chromium, Nickel and Cobalt. In *Trace Elements in Soils*; Hooda, P.S., Ed.; John Wiley & Sons, Ltd.: Chichester, UK, 2010; pp. 461–479, ISBN 978-1-4443-1947-7.

30. Baek, K.; Yang, J.-W. Sorption and desorption of cobalt in clay: Effect of humic acids. *Korean J. Chem. Eng.* **2004**, *21*, 989–993. [CrossRef]

31. Munera-Echeverri, J.; Martinsen, V.; Strand, L.; Zivanovic, V.; Cornelissen, G.; Mulder, J. Cation exchange capacity of biochar: An urgent method modification. *Sci. Total Environ.* **2018**, *642*, 190–197. [CrossRef]

32. Olsen, S.; Cole, C.; Watanabe, F.; Dean, L. *Estimation of Available Phosphorus in Soils by Extraction with Sodium Bicarbonate*; Circular 93; US Department of Agriculture: Washington, DC, USA, 1954.

33. Wang, T.; Camps-Arbestain, M.; Hedley, M.; Bishop, P. Predicting phosphorus bioavailability from high-ash biochars. *Plant Soil* **2012**, *357*, 173–187. [CrossRef]

34. *Standard Reference Standard Test Methods for Loss on Ignition (LOI) of Solid Combustion Residues*; ASTM D7348-13; ASTM International: West Conshohocken, PA, USA, 2013.

35. Knödel, K.; Lange, G.; Voigt, H. *Environmental Geology: Handbook of Field Methods and Case Studies*; Springer: Berlin/Heidelberg, Germany; New York, NY, USA, 2007; ISBN 978-3-540-74669-0.

36. Muyumba, D.; Pourret, O.; Liénard, A.; Bonhoure, J.; Mahy, G.; Luhembwe, M.; Colinet, G. Mobility of copper and cobalt in metalliferous ecosystems: Results of a lysimeter study in the Lubumbashi Region (Democratic Republic of Congo). *J. Geochem. Explor.* **2019**, *196*, 208–218. [CrossRef]

37. *Adsorption—Desorption Using a Batch Equilibrium Method*; OECD Test No. 106; OECD Guidelines for the Testing of Chemicals, Section 1; Organization for Economic Co-operation and Development: Paris, France, 2000.

38. Ćwieląg-Piasecka, I.; Medyńska-Juraszek, A.; Jerzykiewicz, M.; Dębicka, M.; Bekier, J.; Jamroz, E.; Kawałko, D. Humic acid and biochar as specific sorbents of pesticides. *J. Soils Sediments* **2018**, *18*, 2692–2702. [CrossRef]

39. Michalak, I.; Mironiuk, M.; Marycz, K. A comprehensive analysis of biosorption of metal ions by macroalgae using ICP-OES, SEM-EDX and FTIR techniques. *PLoS ONE* **2018**, *13*, e0205590. [CrossRef]

40. Michalak, I.; Chojnacka, K.; Marycz, K. Using ICP-OES and SEM-EDX in biosorption studies. *Microchim. Acta* **2011**, *172*, 65–74. [CrossRef]

41. *FAO World Reference Base for Soil Resources 2014: International Soil Classification System for Naming Soils and Creating Legends for Soil Maps*; FAO: Rome, Italy, 2014; ISBN 978-92-5-108369-7.

42. Gregorich, E.; Carter, M. *Soil Sampling and Methods of Analysis*, 2nd ed.; CRC Press: Boca Raton, FL, USA, 2007; ISBN 9780849335860.

43. Albanese, S.; Sadeghi, M.; Lima, A.; Cicchella, D.; Dinelli, E.; Valera, P.; Falconi, M.; Demetriades, A.; De Vivo, B. GEMAS: Cobalt, Cr, Cu and Ni distribution in agricultural and grazing land soil of Europe. *J. Geochem. Explor.* **2015**, *154*, 81–93. [CrossRef]

44. Ptistišek, N.; Milačič, R.; Veber, M. Use of the BCR three-step sequential extraction procedure for the study of the partitioning of Cd, Pb and Zn in various soil samples. *J. Soils Sediments* **2001**, *1*, 25–29. [CrossRef]

45. Tokalioglu, Ş.; Kartal, Ş.; Elçi, L. Determination of heavy metals and their speciation in lake sediments by flame atomic absorption spectrometry after a four-stage sequential extraction procedure. *Anal. Chim. Acta* **2000**, *413*, 33–40. [CrossRef]

46. Nakamoto, K. *Infrared and Raman Spectra of Inorganic and Coordination Compounds: Part A: Theory and Applications in Inorganic Chemistry*, 6th ed.; Wiley: Hoboken, NJ, USA, 2008; ISBN 978-0-471-74339-2.

47. Palacios, E.; Juárez-López, G.; Monhemius, A. Infrared spectroscopy of metal carboxylates. *Hydrometallurgy* **2004**, *72*, 139–148. [CrossRef]

48. Medyńska-Juraszek, A.; Ćwieląg-Piasecka, I. Effect of Biochar Application on Heavy Metal Mobility in Soils Impacted by Copper Smelting Processes. *Pol. J. Environ. Stud.* **2020**, *29*, 1749–1757. [CrossRef]

49. Bogusz, A.; Oleszczuk, P. Sequential extraction of nickel and zinc in sewage sludge- or biochar/sewage sludge-amended soil. *Sci. Total Environ.* **2018**, *636*, 927–935. [CrossRef]

50. Lehmann, J.; Joseph, S. *Biochar for Environmental Management: Science and Technology*; Earthscan: London, UK, 2009; ISBN 978-84407-658-1.

51. Gul, S.; Whalen, J.; Thomas, B.; Sachdeva, V.; Deng, H. Physico-chemical properties and microbial responses in biochar-amended soils: Mechanisms and future directions. *Agric. Ecosyst. Environ.* **2015**, *206*, 46–59. [CrossRef]

52. Song, Y.; Wang, F.; Bian, Y.; Kengara, F.; Jia, M.; Xie, Z.; Jiang, X. Bioavailability assessment of hexachlorobenzene in soil as affected by wheat straw biochar. *J. Hazard. Mater.* **2012**, *217–218*, 391–397. [CrossRef]

53. O'Toole, A.; Knoth de Zarruk, K.; Steffens, M.; Rasse, D. Characterization, Stability, and Plant Effects of Kiln-Produced Wheat Straw Biochar. *J. Environ. Qual.* **2013**, *42*, 429–436. [CrossRef]

54. Shen, Z.; Zhang, Y.; Jin, F.; McMillan, O.; Al-Tabbaa, A. Qualitative and quantitative characterisation of adsorption mechanisms of lead on four biochars. *Sci. Total Environ.* **2017**, *609*, 1401–1410. [CrossRef]

55. Vaughn, S.; Kenar, J.; Thompson, A.; Peterson, S. Comparison of biochars derived from wood pellets and pelletized wheat straw as replacements for peat in potting substrates. *Ind. Crop. Prod.* **2013**, *51*, 437–443. [CrossRef]

56. Cheng, C.; Lehmann, J.; Thies, J.; Burton, S.; Engelhard, M.H. Oxidation of black carbon by biotic and abiotic processes. *Org. Geochem.* **2006**, *37*, 1477–1488. [CrossRef]

57. Qi, F.; Lamb, D.; Naidu, R.; Bolan, N.; Yan, Y.; Ok, Y.; Rahman, M.; Choppala, G. Cadmium solubility and bioavailability in soils amended with acidic and neutral biochar. *Sci. Total Environ.* **2018**, *610–611*, 1457–1466. [CrossRef]

58. Park, J.; Choppala, G.; Bolan, N.; Chung, J.; Chuasavathi, T. Biochar reduces the bioavailability and phytotoxicity of heavy metals. *Plant Soil* **2011**, *348*, 439–451. [CrossRef]

59. Sanders, J. The effect of pH on the total and free ionic concentrations of mnganese, zinc and cobalt in soil solutions. *J. Soil Sci.* **1983**, *34*, 315–323. [CrossRef]

60. Nagpal, N. *Water Quality Guidelines for Cobalt*; Ministry of Water, Land and Air Protection, Water Protection Section, Water, Air and Climate Change Branch: Victoria, BC, Canada, 2004.

61. Bakkaus, E.; Collins, R.; Morel, J.; Gouget, B. Potential phytoavailability of anthropogenic cobalt in soils as measured by isotope dilution techniques. *Sci. Total Environ.* **2008**, *406*, 108–115. [CrossRef] [PubMed]

62. Liu, B.; Huang, Q.; Su, Y.; Wang, M.; Ma, Y.; Kelly, R. Cobalt accumulation and antioxidant system in pakchois under chemical immobilization in fluvo-aquic soil. *J. Soil Sediments* **2018**, *18*, 669–679. [CrossRef]

63. Yin, D.; Wang, X.; Peng, B.; Tan, C.; Ma, L. Effect of biochar and Fe-biochar on Cd and As mobility and transfer in soil-rice system. *Chemosphere* **2017**, *186*, 928–937. [CrossRef]

64. Zhang, J.; Liu, J.; Liu, R. Effects of pyrolysis temperature and heating time on biochar obtained from the pyrolysis of straw and lignosulfonate. *Bioresour. Technol.* **2015**, *176*, 288–291. [CrossRef]

65. Li, X.; Shen, Q.; Zhang, D.; Mei, X.; Ran, W.; Xu, Y.; Yu, G. Functional Groups Determine Biochar Properties (pH and EC) as Studied by Two-Dimensional 13C NMR Correlation Spectroscopy. *PLoS ONE* **2013**, *8*, e65949. [CrossRef]

66. Januszewicz, K.; Kazimierski, P.; Klein, M.; Kardaś, D.; Łuczak, J. Activated Carbon Produced by Pyrolysis of Waste Wood and Straw for Potential Wastewater Adsorption. *Materials* **2020**, *13*, 2047. [CrossRef]

67. Wang, Q.; Wang, B.; Lee, X.; Lehmann, J.; Gao, B. Sorption and desorption of Pb (II) to biochar as affected by oxidation and pH. *Sci. Total Environ.* **2018**, *634*, 188–194. [CrossRef]

68. Dong, X.; Li, G.; Lin, Q.; Zhao, X. Quantity and quality changes of biochar aged for 5 years in soil under field conditions. *Catena* **2017**, *159*, 136–143. [CrossRef]

69. Mia, S.; Dijkstra, F.; Singh, B. Aging Induced Changes in Biochar's Functionality and Adsorption Behavior for Phosphate and Ammonium. *Environ. Sci. Technol.* **2017**, *51*, 8359–8367. [CrossRef] [PubMed]

70. Uchimiya, M.; Klasson, K.; Wartelle, L.; Lima, I. Influence of soil properties on heavy metal sequestration by biochar amendment: 1. Copper sorption isotherms and the release of cations. *Chemosphere* **2011**, *82*, 1431–1437. [CrossRef]

71. Shen, Z.; Som, A.; Wang, F.; Jin, F.; McMilla, O.; Al-Tabbaa, A. Long-term impact of biochar on the immobilisation of nickel (II) and zinc (II) and the revegetation of a contaminated site. *Sci. Total Environ.* **2016**, *542*, 771–776. [CrossRef]

72. Kołodyńska, D.; Wnetrzak, R.; Leahy, J.; Hayes, M.; Kwapiński, W.; Hubicki, Z. Kinetic and adsorptive characterization of biochar in metal ions removal. *Chem. Eng. J.* **2012**, *197*, 295–305. [CrossRef]

73. Lawrinenko, M.; Liard, D. Anion exchange capacity of biochar. *Green Chem.* **2015**, *17*, 4628–4636. [CrossRef]

74. Bibak, A. Cobalt, copper and manganese adsorption by aluminium and iron oxides and humic acid. *Commun. Soil Sci. Plant Anal.* **1994**, *25*, 3229–3239. [CrossRef]

75. Kanungo, S. Adsorption of cations on hydrous oxides of iron. II. adsorption of Mn, Co, Ni, and Zn onto amorphous FeOOH from simple electrolyte solutions as well as from a complex electrolyte solution resembling seawater in major ion content. *J. Colloid Interface Sci.* **1994**, *162*, 93–102. [CrossRef]

76. Landry, C.; Koretsky, C.; Lund, T.; Schaller, M.; Das, S. Surface complexation modeling of Co (II) adsorption on mixtures of hydrous ferric oxide, quartz and kaolinite. *Geochim. Cosmochim. Acta* **2009**, *73*, 3723–3737. [CrossRef]

77. Woodward, G.; Peacock, C.; Oterofarin, A.; Thompson, O.; Brown, A.; Burke, I. A universal uptake mechanism for cobalt (II) on soil constituents: Ferrihydrite, kaolinite, humic acid, and organo-mineral composites. *Geochim. Cosmochim. Acta* **2018**, *238*, 270–291. [CrossRef]

Article

Composite Properties of Non-Cement Blended Fiber Composites without Alkali Activator

Wei-Ting Lin [1,*], Kae-Long Lin [2], Kinga Korniejenko [3], Lukáš Fiala [4], An Cheng [1] and Jie Chen [1]

[1] Department of Civil Engineering, National Ilan University, No.1, Sec. 1, Shennong Rd., I-Lan 260, Taiwan; ancheng@niu.edu.tw (A.C.); jchen@niu.edu.tw (J.C.)
[2] Department of Environmental Engineering, National Ilan University, No.1, Sec. 1, Shennong Rd., I-Lan 260, Taiwan; kllin@niu.edu.tw
[3] Institute of Materials Engineering, Faculty of Materials Engineering and Physics, Cracow University of Technology, Warszawska 24, 31-155 Kraków, Poland; kkorniej@gmail.com
[4] Department of Materials Engineering and Chemistry, Faculty of Civil Engineering, Czech Technical University in Prague, Thákurova 7, 166 29 Prague 6, Czech Republic; fialal@fsv.cvut.cz
* Correspondence: wtlin@niu.edu.tw; Tel.: +886-3-931-7567

Received: 19 February 2020; Accepted: 20 March 2020; Published: 22 March 2020

Abstract: The vigorous promotion of reuse and recycling activities in Taiwan has solved a number of problems associated with the treatment of industrial waste. Considerable advances have been made in the conversion of waste materials into usable resources, thereby reducing the space required for waste storage and helping to conserve natural resources. This study examined the use of non-alkali activators to create bonded materials. Our aims were to evaluate the feasibility of using ground-granulated blast-furnace slag (S) and circulating fluidized bed co-fired fly ash (F) as non-cement binding materials and determine the optimal mix proportions (including embedded fibers) with the aim of achieving high dimensional stability and good mechanical properties. Under a fixed water/binder ratio of 0.55, we combined S and F to replace 100% of the cement at S:F ratios of 4:6, 5:5, 6:4. Polypropylene fibers (L/d = 375) were also included in the mix at 0.1%, 0.2% and 0.5% of the volume of all bonded materials. Samples were characterized in terms of flowability, compressive strength, tensile strength, water absorption, shrinkage, x-ray diffraction (XRD) and scanning electron microscope (SEM) analysis. Specimens made with an S:F ratio of 6:4 achieved compressive strength of roughly 30 MPa (at 28 days), which is the 80% the strength of conventional cement-based materials (control specimens). The inclusion of 0.2% fibers in the mix further increased compressive strength to 35 MPa and enhanced composite properties.

Keywords: fiber reinforced; cementless composites; microscopic property; co-fired fly ash; green materials

---

1. Introduction

Economic development inevitably increases construction activity, which depends heavily on the production of cement. The most direct approach to reduce the CO_2 emissions associated with the manufacture of cement is to reduce cement consumption or replace cement with other pozzolanic materials (e.g., industrial by-products) with similar binding properties [1–3]. Non-cement blended materials can help to reduce construction costs and the negative impact of cement production on the natural environment [4,5]. Considerable advances have been made in the conversion of waste materials into usable resources [6,7], thereby reducing the space required for waste storage and helping to conserve natural resources [8,9].

Researchers have demonstrated that the incorporation of pozzolans as a partial replacement for cement can improve the mechanical properties and durability of the resulting concrete [10–12].

Numerous studies have also used pozzolans (fly ash or slag) to entirely replace cement in ordinary concrete or mortar. Many specimens have strength, mechanical properties and durability superior to those of conventional concrete [13,14]. The key factor in achieving a fully hydrated reaction without cement is the inclusion of an alkali activator or the enactment of curing at an elevated temperature. The cementless composites containing alkali activators inevitably increase production costs. However, many industrial by-products can be used as alkali activators to save cost due to its chemical composition with higher alkali contents.

Considerable advances have been made in the low-energy manufacture of non-cement blended materials, particularly in Taiwan, Korea and Japan. Ground-granulated blast-furnace slag can be combined with various supplementary cementitious materials to eliminate the need for alkali activators in the production of materials suitable for civil construction. Many such materials provide high compressive strength (30 MPa to 60 MPa), as well as excellent mechanical properties and durability [15–18]. The inclusion of polypropylene fibers in composites has been shown to enhance the tensile strength and volume stability of non-cement blended materials [19–21]. In recent years, a number of researchers have studied the use of polypropylene fibers or non-cement blended materials; however, there has been little work on the addition of polypropylene fibers to non-cement blended materials. In this study, we combined ground-granulated blast-furnace slag (GGBS) with circulating fluidized bed (CFB) co-fired fly ash and polypropylene fibers to produce non-cement blended materials without alkali activator. Samples were compared with standard Portland cement mortar in terms of mechanical properties, permeability and microstructure. The proposed material is intended to be used as a controlled low-strength material, pervious concrete, reinforced recycled concrete and for other engineering applications [22–24].

2. Materials and Methods

2.1. Materials

This study used Type I Portland cement with a specific gravity of 3.15 and fineness of 3310 cm^2/g. The fine aggregate was natural river sand with saturated surface dry (SSD) specific gravity of 2.70, absorption of 1.63% and fineness modulus of 2.33. Hereafter, when used as constituent of non-cement binder, GGBS is referred to as slag (S) and CFB co-fired fly ash is referred to as fly ash (FA). FA was produced in the form of a black powder by Yong Feng Yu, Taiwan with a specific gravity of 2.73. The specific gravity of S was 2.88 and the fineness was 5860 cm^2/g. The particles of FA passing through a No. 100 sieve (150 μm) are roughly 98% and the fineness was 2800 cm^2/g. SEM images of the FA revealed irregular polygonal particles similar to those of S as well as rough surfaces, as shown in Figure 1. Figure 2 presents XRD patterns of FA.

(a)   (b)

Figure 1. SEM images: (**a**) circulating fluidized bed (CFB) co-fired fly ash and (**b**) ground-granulated blast-furnace slag (GGBS).

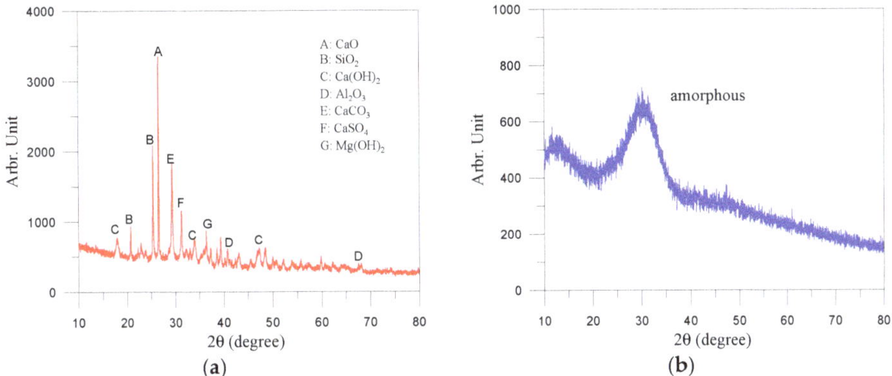

Figure 2. x-ray diffraction (XRD) pattern: (**a**) co-fired fly ash and (**b**) GGBS.

Chemical compositions of FA and S are summarized in Table 1. The chemical composition of FA was analyzed using x-ray fluorescence (XRF) as follows: SiO_2 (29.47%), Al_2O_3 (19.27%) and CaO (35.54%). The polypropylene fibers used in the current study were 12mm in length with an aspect ratio of 375 (d = 32μm) and the fibers were produced by Poplar Co., Ltd. (Taipei, Taiwan). The specific gravity, tensile strength and Young's modulus of the fibers were 0.91, 250 and 3500 MPa, respectively.

Table 1. Chemical compositions of co-fired fly ash and GGBS.

Chemical Compositions	Co-Fired Fly Ash	GGBS
	Content, wt %	
Silicon dioxide (SiO_2)	29.47	33.68
Aluminum oxide (Al_2O_3)	19.27	14.37
Ferric oxide (Fe_2O_3)	3.49	0.29
Calcium oxide (CaO)	35.54	40.24
Magnesium oxide (MgO)	1.82	7.83
Sulphur trioxide (SO_3)	7.36	0.66
others	3.05	2.93

2.2. Mix Design and Test Methods

The water/cementitious ratio (w/c) of the mortar specimens was maintained at a constant 0.55, whereas the cementitious materials/fine aggregate mass ratio was 1:2.75, in accordance with ASTM C109 specifications. Table 2 lists the mix design for non-cement blended materials. The specimens were numbered using two number/letters pairs indicating (1) S:F ratio and (2) the percentage of polypropylene fibers. The number after the letter S indicates the percentage of co-fired fly ash (e.g., S40 means 40% fly ash). P refers to ordinary Portland mortar. The last number indicates the percentage of polypropylene fiber (e.g., F1 means 0.1% fiber). In addition, the mixture for a preliminary test of non-cement blended composites was shown in Table 3. The S, FA, fine aggregates and fibers were mixed for about 10 min by using a low speed mixing machine to create the homogeneous mortar specimens and then cast into the metal molds. After demolding, the specimens for standard curing were placed in the standard curing room until the testing age. Table 4 presents the tests performed, the dimensions of the specimens and the standards used in this study.

Table 2. Mix design of the mortar specimens produced (kg/m^3).

Mix No.	Cement	GGBS	Co-Fired Fly Ash	Fine Aggregates	Water	Superplasticizers	Fiber
P	514	0	0		282	0	
S50		257	257		274.6	7.4	
S40	0	308	206		276.8	5.2	0
S60		206	308		267.2	14.8	
P-F1							0.88
P-F2	514	0	0		282	0	1.76
P-F5							4.41
S50-F1				1412	274.6	7.4	0.88
S50-F2		257	257		274.6	7.4	1.76
S50-F5					270.9	11.1	4.41
S40-F1					276.8	5.2	0.88
S40-F2	0	308	206		274.6	7.4	1.76
S40-F5					270.9	11.1	4.41
S60-F1					267.2	14.8	0.88
S60-F2		206	308		267.2	14.8	1.76
S60-F5					264.2	17.8	4.41

Table 3. Mix design of the mortar specimens for a preliminary test (kg/m^3).

Mix No.	Cement	GGBS	Co-Fired Fly Ash	Fine Aggregates	Water
P	514	0	0		
S10	0	463	51		
S20	0	411	103		
S30	0	360	154	1412	282
S40	0	308	206		
S50	0	257	257		
S60	0	206	308		

Table 4. Test methods.

Test Target		Specimen Dimensions (mm)	Referenced Standard	Testing Age (days)
Fresh properties	flow test	–	ASTM C230	–
Mechanical properties	compressive strength test	50 × 50 × 50	ASTM C109	7, 28, 56
	tensile strength test	Briquet Specimens	ASTM C260 CRD-C 260-01	7, 28
	drying shrinkage test	285 × 25 × 25	ASTM C596	2~28
Permeability	water absorption test	50 × 50 × 50	ASTM C642	56
Microstructure observations	SEM observation	10 × 10 × 3	ASTM C1723	28
	XRD analysis	powders	ASTM C1365	28, 56

3. Results and Discussion

3.1. Flowability

An increase in the proportion of FA significantly decreases the flowability of the non-cement blended materials as shown in Figure 3. As shown in Table 2, note that suitable mixtures (flow of 110% as shown in Figure 3) were obtained only after adding a water-reducing admixture (superplasticizer) expect for the P series specimens. The 110%-line in Figure 3 indicates the referred flowability in accordance with ASTM C109. For each mixture, the superplasticizer was used to control the workability of the specimens. The results indicated that the specimens containing polypropylene fibers had lower flowability. The amount of superplasticizer required to maintain fluidity was proportional to the amount of FA, due to the high absorbency and angularity of the particles (rough surface as shown in Figure 1). In the cement-based composites, flowability decreased with an increase in the proportion of polypropylene fibers, due to the internal resistance and friction generated through the interaction of fibers. The finding is consistent with the reported in previous studies [25].

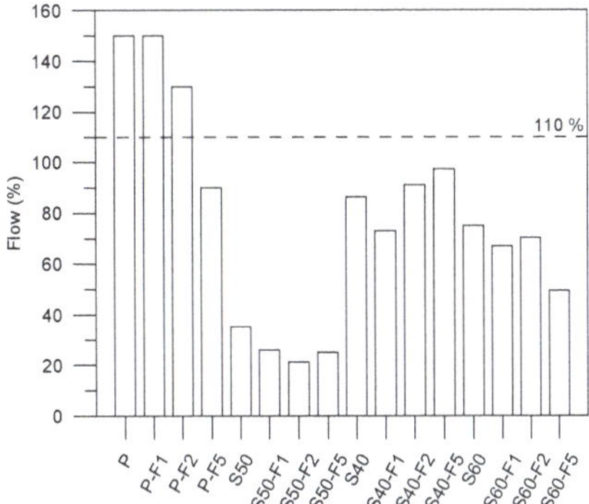

Figure 3. Flow test results.

3.2. Compressive Strength

Figure 4a presents the preliminary test of the compressive strength of specimens made with non-cement materials at various S/FA ratios without an alkali activator. Compressive strength was shown to increase significantly with curing age between 28 and 56 days. The compressive strength of specimens with 40% FA was significantly higher than that of specimens with only 10% FA, at all stages of curing. The compressive strength of the non-cement specimens reached 70% that of the mortar made using Portland cement. The non-cement blended specimens with the highest compressive strength were those made using an S/FA ratio of 6:4. Results of compressive strength development were indicated that S and FA reacted with water and then with calcium hydroxide via a pozzolanic reaction to form hydration products (i.e., calcium silicate hydrate (C-S-H) and/or calcium aluminium silicate hydrate (C-A-S-H) colloids). These findings are consistent with those reported in previous studies [18,24] and it was produced produces viscous hardening behavior in the cementless composites [23].

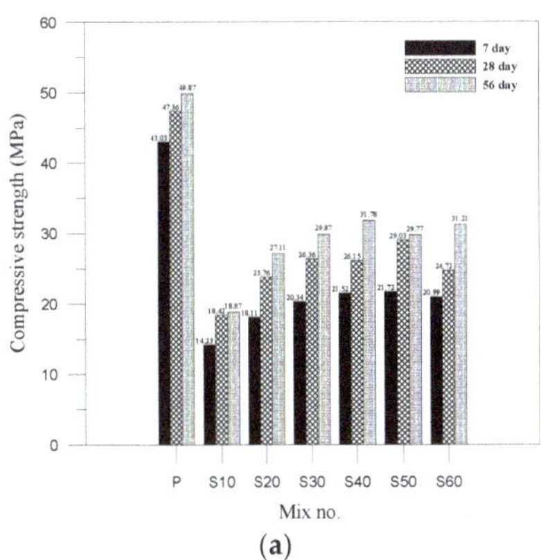
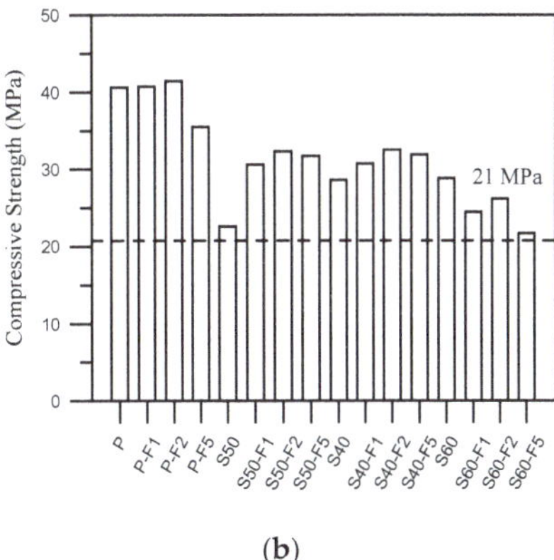

(a)　　　　　　　　　　　　　　　　**(b)**

Figure 4. (a) Histogram presenting compressive strength results of the preliminary test; and (b) compressive strength development at 28 days for all mixes.

Non-cement blended composite without alkali activator is an innovative material. For concrete design, the minimum compressive strengths of concrete for normal structural components such as walls or slabs were used as 21 MPa. For this reason, the target strength was set as 21 MPa in this study. As shown in Figure 4b, all of the mixtures reached the target strength of 21 MPa, which is suitable for normal use in civil and construction. We observed a positive correlation between the proportion of fibers and compressive strength, due perhaps to suppressed crack formation under axial loading. The inclusion of fibers in the non-cement blended composites significantly increased the compressive strength, particularly in S50-F2 and S40-F2 (30% to 40% higher than the S50) [26]. This also had indirect effects on the toughness and mechanical properties of the composites. Hydration reactions among the fibers can increase the interfacial bonding strength, resulting in higher compressive strength. These findings are also consistent with those reported in previous studies [27]. The decrease in compressive strength observed in Figure 4b when the fiber content is increased from mixes P-F2 to P-F5. This diminution may be attributed to the lump of noncontiguous fibers, which is consistent with previous study [28].

3.3. Tensile Strength

Figure 5a illustrates the tensile strength of non-cement blended specimens and cement mortar specimens without fibers. Clearly, the increase in tensile strength in the non-cement mortar was not significant at 7 days compared to the P specimens (average 15% lower than P specimens), but quite noticeable at 28 days (the tensile strength of S60 specimen was slight higher than that of P specimens). The tensile strength of S60 was 18% higher than that of S40. We speculate that this can be attributed to the mix ratio, which allowed the S and FA to be gelatinized; however, the slowness of the reaction hindered strength development. On the basis of the previous study [29], FA can be used as a sustainable alkali activator for S to activate the alkali-activated or pozzolanic reaction. The tensile strength of the non-cement blended specimens increased with an increase in the proportion of polypropylene fibers, as shown in Figure 5b. At 28 days, the tensile strength of samples S50-F2 and S60-F2 were respectively 20% and 23% higher than that of the control specimens. This obvious increase in tensile strength can be attributed to the fibers arresting crack propagation at the micro scale. In addition, the decrease in tensile strength when the fiber content is increased from P-F2, PF5, S50-F2 and S50-F5 was also due to the lump of noncontiguous fibers in composites.

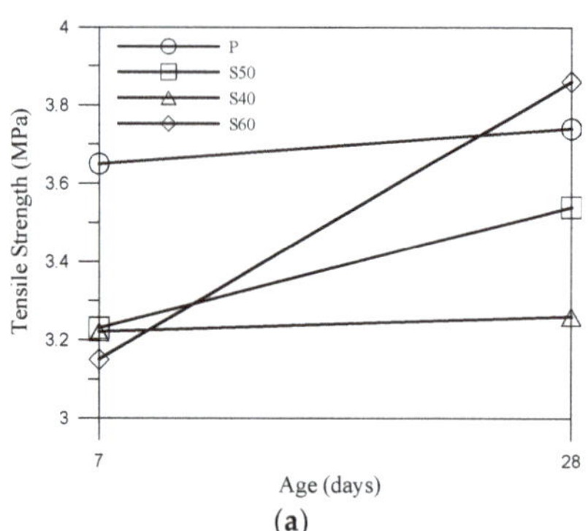

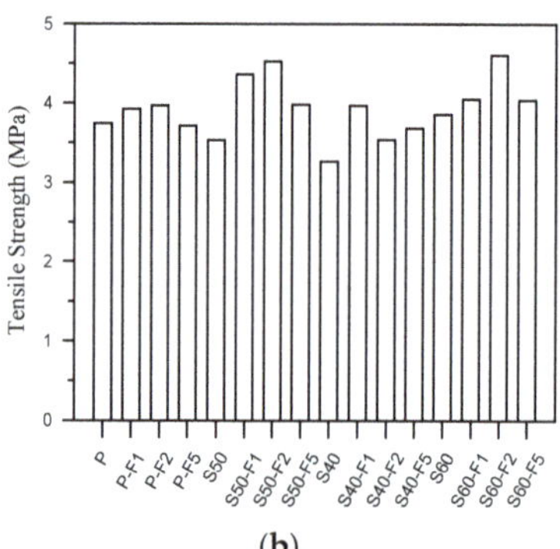

Figure 5. (**a**) Tensile strength development curves of select samples and (**b**) tensile strength results of all samples at 28 days.

3.4. Drying Shrinkage

As shown in Figure 6a, the shrinkage development curves of S40 and S50 specimens were similar to those of P specimens. By contrast, the shrinkage development curve of specimen S60 was 4 times higher than that of cement mortar, indicating a high SO_3 content following a rapid hydration reaction with a corresponding rapid volume expansion. Previous studies [17,18,24] have also reported on the considerable influence of SO$_3$ on the strength of mortar specimens containing FA.

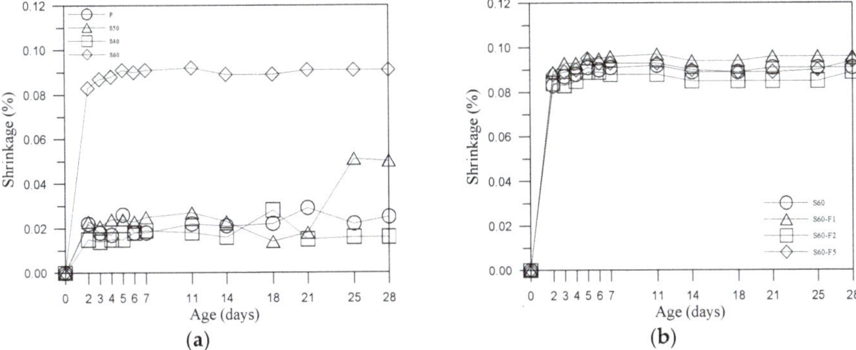

Figure 6. Length change results: (**a**) comparison to P, S40, S50 and S60 specimens and (**b**) comparison to S60, S60-F1, S60-F2 and S60-F5.

Our results revealed that the addition of fibers to specimen S60 (S60-F1/2/5) did not have a significant effect on shrinkage and the effect of fibers in shrinkage development curves of S40 and S50 specimens were similar to those of S60 specimens. We also found that shrinkage was largely independent of the proportion of FA. FA contains large quantities of free lime (f-CaO) and SO$_3$, which are beneficial to strength development of hydrated in non-cement blended composites containing S. Note however that f-CaO and/or excessive SO$_3$ can cause expansion in hardened mortar. Increasing the amount of FA could increase the number of sulphate ions in solution, thereby hindering ettringite formation [29,30]. This can have profound effects on the volume stability and strength of mortar specimens with a high proportion of FA.

3.5. Absorption

Figure 7 presents the water absorption results of all mixtures. The absorption of S40 specimens was lower than that of the P, S50 and S60. The results indicated that the absorption decreased with an increase in the proportion of S due changes in the pore structure filled by finer S. The inclusion of fine S powder improved the compactness through pozzolanic reactivity and pore filling effects. Absorption also decreased with an increase in the proportion of fibers, due to the fact that the fibers blocked the connectivity of pores, thereby making the transmission paths more complex.

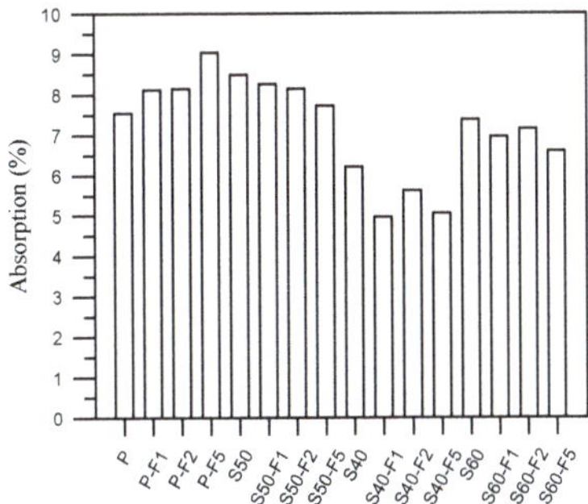

Figure 7. Absorption results of all samples at 56 days.

3.6. XRD Analysis

Figure 8 presents XRD patterns of non-cement blended samples prepared using various S/FA ratios at 28 days. Hydration products appeared in various phases, including $Ca(OH)_2$, Ca_3SiO_5, $Ca_6Al_2(OH)_{12}(SO_4)_3$-$26H_2O$ and $CaFe_2O_5$. The non-cement blended specimens displayed roughly the same $CaSO_4$-$2H_2O$ and $Ca_6Al_2(OH)_{12}(SO_4)_3$-$26H_2O$ (ettringite) [31] peaks compared to the control groups. Note that the non-cement blended specimens also presented relatively small $Ca(OH)_2$ peaks and slightly high Ca_3SiO_5 peaks compared to the P specimens. It also indicated that the hydration products of non-cement blended specimens were similar to the P specimens and almost less calcium hydroxide ($Ca(OH)_2$) was found in non-cement blended specimens. It indicated that the quartz (SiO_2) in the FA reacted with lime (CaO) and $Ca(OH)_2$, resulting in the formation of C-S-H and C-A-S-H gel, which are the key factors in strength development in cementitious materials [32,33]. The FA also had higher proportions of f-CaO and $CaSO_4$, which has been shown to activate cementing processes and hydration reactions [34]. It also confirmed that the strength of non-cement blended composites can be achieved to conventional cement-based materials at the age of 56 days.

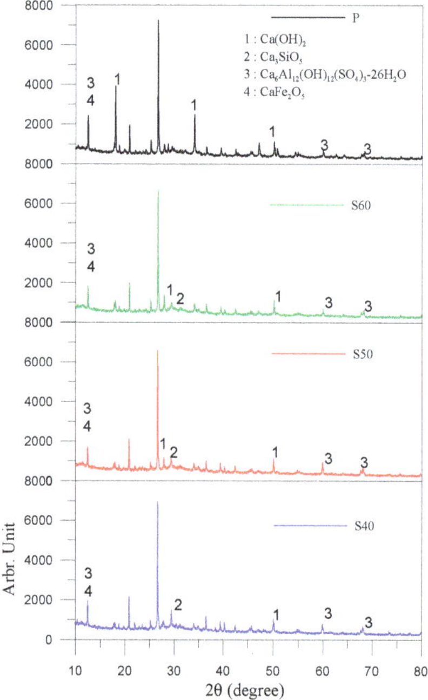

Figure 8. XRD patterns of various samples at 28 days.

3.7. SEM Observation

Figure 9 presents SEM images of various specimens under 10K× magnification at 28 days. The non-cement blended specimens presented more pronounced pore formation than did the cement specimens. C-S-H gel is the product of reactions between water and tricalcium silicate or dicalcium silicate. Approximately 50% of the cement gel was C-S-H gel, which served as the primary source of strength in the cement paste. C-S-H gel and ettringite appeared as irregular needle-like and spherical continuums, whereas $Ca(OH)_2$ appears as hexagonal flakes. However, the appearance of fine exterior capillary tube spikes indicates that the C-S-H gel was covered in pores, i.e., it was not a smooth continuum. The formation of high-quality concrete depends on finer materials filling in the pores.

As shown in the Figures 9 and 10, most of the pores have been filled with ettringite, which gave the material a dense appearance and reduced the likelihood of infiltration by harmful substances.

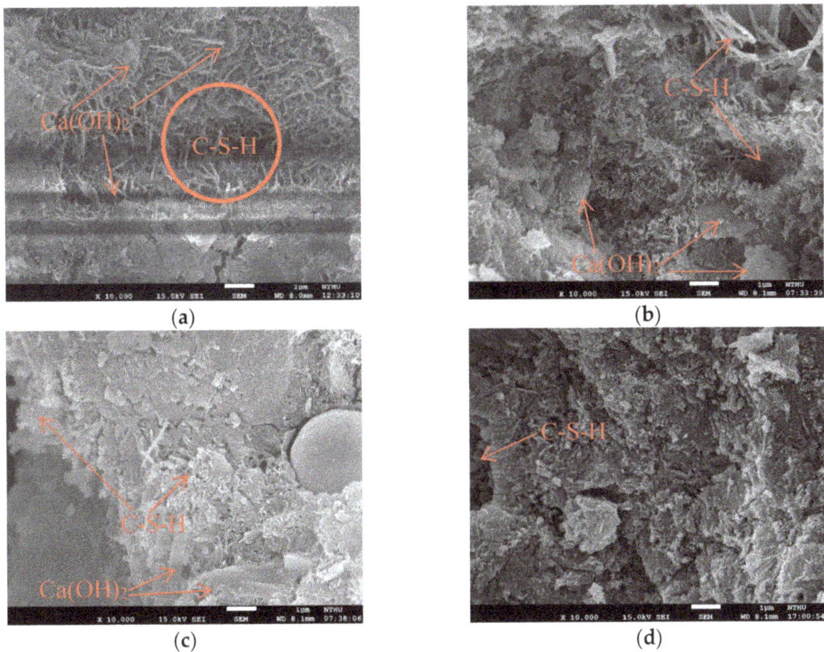

Figure 9. SEM photos at 28 days: (**a**) P; (**b**) S40; (**c**) S50 and (**d**) S60.

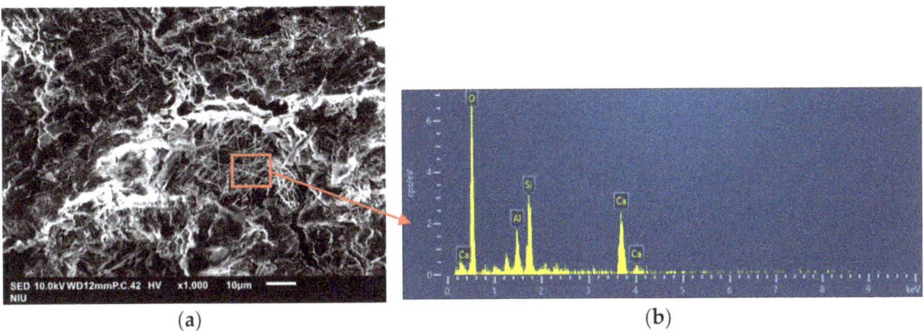

Figure 10. Scanning electron microscope (SEM) photos with energy dispersive analysis (EDS) analysis: (**a**) S60 specimens; (**b**) EDS results.

Based on the SEM images, the reaction mechanisms in non-cement blended materials comprising a mix of S and FA can be divided into two phases. In the first phase, the CaO in the FA violently reacted with water to produce $Ca(OH)_2$. It was shown that increasing the proportion of FA accelerated the process of setting. Insufficient FA would slow the hydration setting speed after the S and FA mix. In the second phase, the $Ca(OH)_2$ and SiO_2 in the S and FA with reacted with SiO_2 and Al_2O_3 to produce C-S-H and C-A-S-H gel [35] as shown in Figure 10. It can be consistent with previous study [18].

4. Conclusions

It is a feasibility to produce entirely non-cement blended materials using S with FA and FA can be as an alkali activator for S in blended composites. In this study, we produced non-cement blended materials by replacing 100% of the cement (by weight) with S and FA at ratios of 4:6, 5:5 and 6:4. Specimens made with an S:FA ratio of 6:4 achieved compressive strength of roughly 30 MPa (at 28 days), which is the 80% the strength of conventional cement-based materials (control specimens). Non-cement blended composites were used to have a proper inclusion of fiber reinforcements to conduct the engineering properties and increase its applicability for construction applications. The inclusion of 0.2% fibers in the mix further increased compressive strength to 35 MPa. It also enhanced the compactness of micro pore-structures, increased the compressive strength and tensile strength and decreased absorption and the likelihood of shrinkage. SEM images and XRD analysis revealed that the compressive strength of the non-cement blended specimens can be attributed to the formation of C-S-H and C-A-S-H gels by $Ca(OH)_2$, SiO_2, and Al_2O_3. Polypropylene fibers were shown to have a profound effect in reinforcing the non-cement blended materials and it has shown the feasibility of the usage of non-cement blended fiber composites in civil construction.

Author Contributions: Methodology, W.-T.L. and A.C.; validation, K.K. and L.F.; data curation, J.C. and W.-T.L.; writing—original draft preparation, W.-T.L.; writing—review and editing, W.-T.L.; visualization, K.-L.L. All authors have read and agreed to the published version of the manuscript.

Funding: This research was funded by Ministry of Science and Technology (MOST) under the Grant MOST 108-2221-E-197-006 in Taiwan and the Czech Science Foundation under the project No. 19-11516S.

Acknowledgments: This research was acknowledged by the Polish National Agency for Academic Exchange under the International Academic Partnership Programme within the framework of the grant: E-mobility and sustainable materials and technologies EMMAT.

Conflicts of Interest: The authors declare no conflict of interest.

References

1. Wang, J.; Wu, H.; Duan, H.; Zillante, G.; Zuo, J.; Yuan, H. Combining life cycle assessment and Building Information Modelling to account for carbon emission of building demolition waste: A case study. *J. Clean. Prod.* **2018**, *172*, 3154–3166. [CrossRef]
2. Koytsoumpa, E.I.; Bergins, C.; Kakaras, E. The CO_2 economy: Review of CO_2 capture and reuse technologies. *J. Supercrit. Fluids* **2018**, *132*, 3–16. [CrossRef]
3. Liguori, B.; Iucolano, F.; De Gennaro, B.; Marroccoli, M.; Caputo, D. Zeolitized tuff in environmental friendly production of cementitious material: Chemical and mechanical characterization. *Constr. Build. Mater.* **2015**, *99*, 272–278. [CrossRef]
4. Lu, W.; Webster, C.; Chen, K.; Zhang, X.; Chen, X. Computational Building Information Modelling for construction waste management: Moving from rhetoric to reality. *Renew. Sustain. Energy Rev.* **2017**, *68*, 587–595. [CrossRef]
5. Sapuay, S. Construction Waste—Potentials and Constraints. *Procedia Environ. Sci.* **2016**, *35*, 714–722. [CrossRef]
6. Ferone, C.; Capasso, I.; Bonati, A.; Roviello, G.; Montagnaro, F.; Santoro, L.; Turco, R.; Cioffi, R. Sustainable management of water potabilization sludge by means of geopolymers production. *J. Clean. Prod.* **2019**, *229*, 1–9. [CrossRef]
7. Capasso, I.; Lirer, S.; Flora, A.; Ferone, C.; Cioffi, R.; Caputo, D.; Liguori, B. Reuse of mining waste as aggregates in fly ash-based geopolymers. *J. Clean. Prod.* **2019**, *220*, 65–73. [CrossRef]
8. Karim, M.; Hossain, M.; Zain, M.; Jamil, M.; Lai, F. Durability properties of a non-cement binder made up of pozzolans with sodium hydroxide. *Constr. Build. Mater.* **2017**, *138*, 174–184. [CrossRef]
9. Hemalatha, M.; Santhanam, M. Characterizing supplementary cementing materials in blended mortars. *Constr. Build. Mater.* **2018**, *191*, 440–459. [CrossRef]
10. Divsholi, B.S.; Lim, T.Y.D.; Teng, S. Durability Properties and Microstructure of Ground Granulated Blast Furnace Slag Cement Concrete. *Int. J. Concr. Struct. Mater.* **2014**, *8*, 157–164. [CrossRef]

11. Yahaya, F.M.; Muthusamy, K.; Sulaiman, N. Corrosion Resistance of High Strength Concrete Containing Palm Oil Fuel Ash as Partial Cement Replacement. *Res. J. Appl. Sci. Eng. Technol.* **2014**, *7*, 4720–4722. [CrossRef]

12. Özen, S.; Goncuoglu, M.; Liguori, B.; De Gennaro, B.; Cappelletti, P.; Gatta, G.D.; Iucolano, F.; Colella, C. A comprehensive evaluation of sedimentary zeolites from Turkey as pozzolanic addition of cement- and lime-based binders. *Constr. Build. Mater.* **2016**, *105*, 46–61. [CrossRef]

13. Hardjito, D.; Wallah, S.E.; Sumajouw, D.M.J.; Rangan, B.V. On the development of fly ash-based geopolymer concrete. *ACI Mater. J.* **2004**, *101*, 467–472.

14. Chi, M.; Huang, R. Binding mechanism and properties of alkali-activated fly ash/slag mortars. *Constr. Build. Mater.* **2013**, *40*, 291–298. [CrossRef]

15. Zhang, W.; Choi, H.; Sagawa, T.; Hama, Y. Compressive strength development and durability of an environmental load-reduction material manufactured using circulating fluidized bed ash and blast-furnace slag. *Constr. Build. Mater.* **2017**, *146*, 102–113. [CrossRef]

16. Nguyen, H.-A.; Chang, T.-P.; Shih, J.-Y.; Chen, C.-T.; Nguyen, T.-D. Influence of circulating fluidized bed combustion (CFBC) fly ash on properties of modified high volume low calcium fly ash (HVFA) cement paste. *Constr. Build. Mater.* **2015**, *91*, 208–215. [CrossRef]

17. Chi, M. Synthesis and characterization of mortars with circulating fluidized bed combustion fly ash and ground granulated blast-furnace slag. *Constr. Build. Mater.* **2016**, *123*, 565–573. [CrossRef]

18. Wu, Y.-H.; Huang, R.; Tsai, C.-J.; Lin, W.-T. Utilizing residues of CFB co-combustion of coal, sludge and TDF as an alkali activator in eco-binder. *Constr. Build. Mater.* **2015**, *80*, 69–75. [CrossRef]

19. Guo, H.; Tao, J.; Chen, Y.; Li, D.; Jia, B.; Zhai, Y. Effect of steel and polypropylene fibers on the quasi-static and dynamic splitting tensile properties of high-strength concrete. *Constr. Build. Mater.* **2019**, *224*, 504–514. [CrossRef]

20. Shen, D.; Liu, X.; Zeng, X.; Zhao, X.; Jiang, G. Effect of polypropylene plastic fibers length on cracking resistance of high performance concrete at early age. *Constr. Build. Mater.* **2020**, *244*, 117874. [CrossRef]

21. Szelag, M.; Szelag, M. Evaluation of cracking patterns of cement paste containing polypropylene fibers. *Compos. Struct.* **2019**, *220*, 402–411. [CrossRef]

22. Lin, W.-T.; Weng, T.-L.; Cheng, A.; Chao, S.-J.; Hsu, H.-M. Properties of Controlled Low Strength Material with Circulating Fluidized Bed Combustion Ash and Recycled Aggregates. *Materials* **2018**, *11*, 715. [CrossRef] [PubMed]

23. Ho, H.-L.; Huang, R.; Hwang, L.-C.; Lin, W.-T.; Hsu, H.-M. Waste-Based Pervious Concrete for Climate-Resilient Pavements. *Materials* **2018**, *11*, 900. [CrossRef] [PubMed]

24. Lin, W.-T.; Lin, K.; Chen, K.; Korniejenko, K.; Hebda, M.; Łach, M. Circulation Fluidized Bed Combustion Fly Ash as Partial Replacement of Fine Aggregates in Roller Compacted Concrete. *Materials* **2019**, *12*, 4204. [CrossRef]

25. Kuder, K.G.; Ozyurt, N.; Mu, E.B.; Shah, S.; Özyurt, N. Rheology of fiber-reinforced cementitious materials. *Cem. Concr. Res.* **2007**, *37*, 191–199. [CrossRef]

26. Verdolotti, L.; Iucolano, F.; Capasso, I.; Lavorgna, M.; Iannace, S.; Liguori, B. Recycling and recovery of PE-PP-PET-based fiber polymeric wastes as aggregate replacement in lightweight mortar: Evaluation of environmental friendly application. *Environ. Prog. Sustain. Energy* **2014**, *33*, 1445–1451. [CrossRef]

27. Han, T.-Y.; Lin, W.-T.; Cheng, A.; Huang, R.; Huang, C.-C. Influence of polyolefin fibers on the engineering properties of cement-based composites containing silica fume. *Mater. Des.* **2012**, *37*, 569–576. [CrossRef]

28. Mashrei, M.A.; Sultan, A.; Mahdi, A.M. Effects of polypropylene fibers on compressive and flexural strength of concrete material. *Int. J. Civ. Eng. Technol.* **2018**, *9*, 2208–2217.

29. Wu, Y.-H.; Huang, R.; Tsai, C.-J.; Lin, W.-T. Recycling of Sustainable Co-Firing Fly Ashes as an Alkali Activator for GGBS in Blended Cements. *Materials* **2015**, *8*, 784–798. [CrossRef]

30. Sheng, G.; Zhai, J.; Li, Q.; Li, F. Utilization of fly ash coming from a CFBC boiler co-firing coal and petroleum coke in Portland cement. *Fuel* **2007**, *86*, 2625–2631. [CrossRef]

31. Guo, B.; Xiong, Y.; Chen, W.; Saslow, S.A.; Kozai, N.; Ohnuki, T.; Dabo, I.; Sasaki, K. Spectroscopic and first-principles investigations of iodine species incorporation into ettringite: Implications for iodine migration in cement waste forms. *J. Hazard. Mater.* **2020**, *389*, 121880. [CrossRef]

32. Anthony, E.J.; Jia, L.; Wu, Y. CFBC ash hydration studies. *Fuel* **2005**, *84*, 1393–1397. [CrossRef]

33. Lin, K.; Cheng, T.-W.; Ho, C.-H.; Chang, Y.-M.; Lo, K.-W. Utilization of Circulating Fluidized Bed Fly Ash as Pozzolanic Material. *Open Civ. Eng. J.* **2017**, *11*, 176–186. [CrossRef]

34. Poon, C.; Kou, S.; Lam, L.; Lin, Z. Activation of fly ash/cement systems using calcium sulfate anhydrite (CaSO4). *Cem. Concr. Res.* **2001**, *31*, 873–881. [CrossRef]
35. Baek, C.; Seo, J.; Choi, M.; Cho, J.; Ahn, J.; Cho, K. Utilization of CFBC Fly Ash as a Binder to Produce In-Furnace Desulfurization Sorbent. *Sustainability* **2018**, *10*, 4854. [CrossRef]

 *materials*

Article

Egg By-Products as a Tool to Remove Direct Blue 78 Dye from Wastewater: Kinetic, Equilibrium Modeling, Thermodynamics and Desorption Properties

Ainoa Murcia-Salvador, José A. Pellicer, María Isabel Rodríguez-López, Vicente Manuel Gómez-López, Estrella Núñez-Delicado and José A. Gabaldón *

Dpto. de Ciencias de la Salud., Universidad Católica San Antonio de Murcia (UCAM), Avenida de los Jerónimos s/n, 30107 Murcia, Spain; amurcia6@alu.ucam.edu (A.M.-S.); japellicer@ucam.edu (J.A.P.); mirodriguez@ucam.edu (M.I.R.-L.); vmgomez@ucam.edu (V.M.G.-L.); enunez@ucam.edu (E.N.-D.)

* Correspondence: jagabaldon@ucam.edu; Tel.: +34-968-278-622

Received: 18 February 2020; Accepted: 8 March 2020; Published: 11 March 2020

Abstract: Eggshell, a waste material from food manufacturing, can be used as a potential ecofriendly adsorbent for the elimination of textile dyes from water solutions. The adsorption process was evaluated varying factors such as initial dye load, contact time, pH, quantity of adsorbent, and temperature. The initial dye load (Direct Blue 78) was in the range of 25–300 mg/L. The kinetics of adsorption were analyzed using different models, such as pseudo-first-order, pseudo-second-order, and intraparticle diffusion model. Also, the experimental data at equilibrium were studied using Freundlich, Langmuir, and Temkin isotherms. The kinetics followed pseudo-second-order, then pseudo-first-order, and finally the model of intraparticle diffusion. The results obtained for data at equilibrium follow the order: Freundlich > Langmuir > Temkin. The adsorption equilibrium showed a maximum capacity of adsorption (q_{max}) of 13 mg/g at pH 5, and using 0.5 g of eggshell. Dye adsorption was enhanced with increasing temperatures. The thermodynamic study revealed the spontaneity and endothermic nature of the adsorption process. The desorption study shows that the eggshell could be reused in different adsorption/desorption cycles. A novel advanced oxidation process could degrade more than 95% of the dye. The results show that eggshell is a waste material useful to remove hazardous dyes from wastewater, which may alleviate the environmental impact of dyeing industries.

Keywords: adsorption; eggshell; Direct Blue 78; kinetics; isotherms; pulsed light

1. Introduction

Wastewater polluted with large amounts of dyes is usually dumped into aqueous effluents from the food and plastic, leather, paper, printing, and textile industries [1]. Dyes have been widely used in many industries for coloration purposes thanks to their good features, such as easy application, low production cost, bright colour, and water-fastness [2]. Nowadays, more than 100,000 commercial dyes exist with over 7×10^5 tons of dyes produced per year, and approximately 10–15% are discharged from textile industries [3]. Dyes are organic compounds that are classified according to the chemical composition and type of application. Commercial dyes are categorized into three classes: cationic, anionic, and nonionic, according to the charge after its dissolution in water. Azo compounds are the most typical dyes used for industrial purposes [4–6].

Dyes are non-biodegradable, stable, oxidizing agents and are highly toxic and mutagenic to aquatic life and humans [7]. Their discharge may produce severe hazard to aquatic living organisms, affecting different processes of aquatic vegetation, decreasing the oxygen levels in water and resulting in the choking of aquatic fauna and flora [8]. Thus, it is essential to reduce and remove organics pollutants from wastewater before discharging them [9]. Various methods have been applied for the removal of harmful contaminants from water and wastewater, involving adsorption on activated carbons, reverse osmosis, chemical oxidation, membrane filtration, bacterial action, coagulation and flocculation, activated sludge, ozonation, precipitation, electro-dialysis, ion exchange, and electrochemical techniques [10,11]. However, many of these methods are expensive and exhibit other drawbacks such as poor effectiveness and formation of sludge excess, and are thus unsuitable to be applied by small-scale industries [1]. Between them, adsorption is considered the most feasible method and has become one of the effective and easiest physico-chemical treatment procedures for the decolourization of textile wastewater. It offers several advantages such as low initial costs, high efficiency, producing nontoxic intermediates or by-products, high removal capability, versatility, easy handling, fast adsorption rate, and facile separation [2,12].

The adsorption based on activated carbon is widely employed to remove dyes, however, it still remains an expensive process owing to its high initial cost and the need for a regeneration system of the adsorbent that make it less economically viable [8]. Recently, more effective and cheaper adsorbents based on by-products from poultry waste, agricultural waste, and other natural waste have been developed as an alternative to conventional wastewater treatment processes [13].

In the last decades, researchers have been paying attention to the adsorption properties of agricultural wastes to develop new biosorbents in accordance with circular economy principles. Some alternatives include the orange peel, chitosan, eggshell, grape pomace, coffee residues, rice straw, olive stones, banana peel, artichoke agrowaste, sugarcane bagasse, and bamboo shell, among others [14–24]. This waste, which is eco-friendly, economic renewable, and available in abundance, is a candidate for the treatment of polluted water and wastewater [11]. Among the agricultural poultry by-products, the waste generated can be a promising biosorbent that has slightly increased in recent years. Moreover, the use of by-products from the agro-food sectors in the adsorption process helps to reduce the waste, as well as the low cost biomasses acquisition [17]. Waste materials obtained from different sources have been used as potential adsorbents for the removal of inorganic and organic pollutants. Eggshell and eggshell membranes are waste materials produced in large amounts in the poultry and farms industries as well as restaurants, bakeries, or homes [25].

Approximately 10% of the total mass of hen egg corresponds to eggshell by-product, with an average weigh of 60 g, and it is usually discarded in landfills without any pretreatment. In addition, this waste is commonly reused as soil conditioner, fertilizer, or additive for animal feed. The eggshell consists of three layers: the cuticle on the outer surface (mucin protein), the calcified eggshell (calcite or calcium carbonate crystals), and the eggshell membrane (protein fibers). Numerous pore channels are distributed on the surface of eggshell (between 7000 and 17,000 pores), thus eliciting water transpiration and gaseous exchanges. The porous nature of eggshell makes it a desirable material to be used as an adsorbent [11,13].

Retention of contaminants by adsorbents is not 100% efficient; therefore, additional methods are required to avoid non-retained dyes being disposed of into the environment. The degradation of dyes by an advanced oxidation process (AOP), using pulsed light technology as a photolyzer of hydrogen peroxide, is a novel, fast, and efficient version of AOPs that uses mercury-free lamps to generate energy-dense light [26].

Thus, the main objective of this paper was to evaluate the potential use of eggshell to remove Direct Blue 78 (DB78) dye in aqueous solution by adsorption, and to determine the effect of different parameters such as contact time, initial dye load, pH, and adsorbent concentration. The isotherms of adsorption, as well as the kinetics of dye adsorption on the eggshell, were evaluated by fitting the experimental data to different kinetics and isotherms models. Further, the efficiency of degradation of DB78 by a novel AOP in order to minimize dye discharge to the environment was measured.

2. Materials and Methods

2.1. Materials

Hen eggshells were purchased from a local organic farm. Sodium hydroxide (NaOH), acetic acid ($C_2H_4O_2$), sodium phosphate monobasic monohydrate (H_4NaO_5P), hydrogen peroxide (H_2O_2), and hydrochloric acid (HCl) were obtained from Sigma-Aldrich (Madrid, Spain) and boric acid (H_3BO_3) was purchased from Scharlau (Madrid, Spain). All of these chemicals were used to prepare dye solutions at different pH values. Colorprint (Valencia, Spain) kindly provided Direct Blue 78 dye. Deionized water was utilized to prepare all aqueous solutions throughout the experiments.

2.2. Eggshell Conditioning

Eggshells with their membrane were washed under tap water, dried at room temperature, and stored at $-35\,°C$ in order to prevent the spoilage of the eggshell samples before using them. Before using them, eggshells were ground in a sturdy vessel in which the material was pounded with a pestle and sieved to obtain the required particle size (1–$1.5\ cm^2$). Thereafter, pieces of eggshell were used as the adsorbent in the adsorption studies.

The major constituents of the eggshell are carbonates, sulphates and phosphates of calcium and magnesium, and organic matter. Traces of Na, K, Mn, Fe, Cu, and Sr metals are also present in the eggshell [27]. The density of the eggshell is about 2.53 g/cm^3, which is significantly larger than that of eggshell membrane (1.358 g/cm^3). The major constituents of the eggshell are calcium carbonate (94%), organic matter (4%), calcium phosphate (1%), and magnesium carbonate (1%). The eggshell membrane possesses nearly 60% protein (collagen (35%), glucosamine (10%), chondroitin (9%), and hyaluronic acid (5%)), along with other inorganic components like Ca, Mg, Si, Zn, and so on in smaller quantities [28]. The membrane surface bears positively charged sites produced by basic side chains of amino acids. It has a very high surface area with special functional groups such as hydroxyl ($-OH$), thiol ($-SH$), carboxyl ($-COOH$), amino ($-NH_2$), amide ($-CONH_2$), and so on, which strongly interact with some chemical species present in the albumen.

2.3. Dye Solution Preparation

DB78 is an azo dye (CAS 2503-73-3) whose molecular weight is 1055.91 g/mol and whose formula is $C_{42}H_{25}N_7Na_4O_{13}S_4$. Aqueous solutions with different concentrations of dye (25, 50, 100, 150, 200, 250, and 300 mg/L) were prepared from a 1 g/L stock solution, and used to determine the capacity of adsorption of the eggshell.

2.4. Analyses and Data Evaluation

The concentration of DB78 was measured at the wavelength of 612 nm, which is the maximum absorbance of the dye [15]. A spectrophotometer was used to determine the absorbance before and after the treatment (UV-1603, Shimadzu, Kyoto, Japan).

2.5. Adsorption Experiments

The adsorption experiments were conducted at different dye concentrations, ranging from 25 to 300 mg/L. The kinetic adsorption studies were performed at different doses of adsorbent, pH, and dye concentrations. The effect of contact time on the adsorption capacity of eggshell was conducted while varying dye concentrations from 25 to 300 mg/L. Furthermore, the effect of adsorbent dosage was studied to optimize the adsorption process. pH was carefully adjusted between 3 and 11 to determine the optimum pH to ensure the maximum DB78 removal. Apart from that, the effect of temperature on the adsorption of DB78 by eggshell was investigated at three different temperatures (29, 55, and 75 °C).

A typical experiment was conducted by adding the eggshell to 40 mL of different dye concentrations of the DB78 solution. The mixture was stirred for a predetermined period of time at a constant speed of 500 rpm. The residual dye concentration in the solution was measured at time intervals (20, 40, 60, 80, 100, 120, and 140 min) until the equilibrium conditions were attained. At the end of each interval, the samples were centrifuged at 3000 rpm for 5 min to separate the solid phase, completely removing impurities that may later affect the measure.

The amount of dye adsorbed on eggshell (q_t) at time (t), in mg/g, was determined by Equation (1), as follows [29]:

$$q_t = \frac{V\,(C_0 - C_e)}{m} \tag{1}$$

where V is the volume of dye solution (L); C_0 and C_e are the initial and equilibrium concentrations of dye in liquid phase (mg/L), respectively; and m is the mass of eggshell (g). All the experiments were carried out in triplicate. Three adsorption kinetics models, three isotherms, and the thermodynamic study were evaluated to elucidate the mechanism of dye adsorption.

2.6. Adsorption Kinetics

In order to analyze the mechanism of dye adsorption onto eggshell, and to predict the rate at which a solute (dye) was removed from aqueous solution, three kinetic models could be employed: pseudo-first-order kinetic model [30], pseudo-second-order kinetic model [31], and intraparticle diffusion model [32].

The Lagergren's equation for pseudo-first-order kinetics is given by the following Equation (2) [30]:

$$\log(q_e - q_t) = \log q_e - \frac{k_1}{2.303}t \tag{2}$$

where q_e and q_t are the adsorption capacity at equilibrium and at time (t) (mg/g), respectively, and k_1 (min^{-1}) is the rate constant of this model. From the linear plot of $\log\,(q_e - q_t)$ versus t, the rate constant (k_1) can be obtained by the slope.

The linear form of the Ho and McKay rate equation for pseudo-second-order kinetics is expressed as Equation (3), as follows:

$$\frac{t}{q_t} = \frac{1}{k_2 q_e{}^2} + \frac{1}{q_e}t \tag{3}$$

where q_e and q_t are the adsorption capacity (mg/g) at equilibrium and at time (t), respectively, and k_2 (g/mg min) is the rate constant of this model and can be obtained from the intercept and slope of plot t/q_t versus t [31].

In the adsorption experiments, it is mandatory to fit the experimental data to the intraparticle diffusion model in order to analyze in depth the adsorption behavior of DB78 on eggshell, and to know the rate-determining step in the liquid adsorption systems. In the adsorption of pollutants onto adsorbents, different stages are differentiated; in the first one, there is a transport of the dye from the solution to the adsorbent surface, followed by the diffusion into the adsorbent, which is usually a slow process [32]. In the diffusion model proposed by Weber and Morris [33], the rate can be expressed in terms of the square root time and can be determined as Equation (4), as follows:

$$q_t = k_i \sqrt{t} + C \tag{4}$$

where q_t is the adsorption capacity at any time t (mg/g); and k_i is the rate constant of this model (mg/g min$^{1/2}$) and its values can be calculated from the slopes of plots q_t versus $t^{1/2}$, where t is the time (min) and C is the intercept (mg/g).

2.7. Isotherm Analysis

The interaction between dyes and the adsorbent materials is described using different theoretical models, known as adsorption isotherms. These isotherms are essential in the optimization of the adsorption process [34,35]. Equilibrium isotherm equations were used to describe experimental sorption data [36] and the parameters of equilibrium isotherms provide useful information on adsorption mechanisms, affinity of the adsorbent, and surface properties [20]. Different isotherms were used to analyze the adsorption equilibrium in this study: The Freundlich, Langmuir and Temkin models.

The Freundlich isotherm model suggests heterogeneity in the adsorption sites and takes into account that the adsorption occurs at sites with different energy of adsorption. This isotherm is obtained from the linear form of the Freundlich expression Equation (5) [36]:

$$ln\, q_e = \ln K_F + \frac{1}{n_F} \ln C_e \qquad (5)$$

where C_e (mg/L) and q_e (mg/g) are the liquid phase concentration and solid phase concentration of dye at equilibrium, respectively; $1/n_F$ is the heterogeneity factor; and K_F is the Freundlich constant (L/g) related to the bonding energy. $1/n_F$ and K_F values were calculated from the slope and intercept of plots $\ln q_e$ versus $\ln C_e$, respectively. The values of $1/n_F$ indicate the type of adsorption process: favorable ($0 < 1/n_F < 1$), unfavorable ($1/n_F > 1$), or irreversible ($1/n_F = 0$) [37].

The Langmuir isotherm model assumes that the adsorption process happens at specific homogeneous sites on the adsorbent. This model is probably the most employed adsorption isotherm and is used successfully for the adsorption of contaminants from water solutions. The linearized form of Langmuir model can be given as follows Equation (6) [33,38]:

$$\frac{C_e}{q_e} = \frac{1}{K_L} + \frac{a_L}{K_L} C_e \qquad (6)$$

where C_e is the dye concentration at equilibrium in solution (mg/L), q_e is the adsorption capacity (mg/g) at equilibrium time, and K_L (L/g) and a_L (L/mg) are the Langmuir isotherm constants. The constants K_L and a_L can be calculated from the intercept ($1/K_L$) and the slope (a_L/K_L) of the linear plot between C_e/q_e, and C_e. q_{max} is the maximum adsorption capacity of adsorbent (mg/g) and is determined by K_L/a_L.

The fundamental characteristic of Langmuir isotherm is the separation factor, which is a dimensionless constant (R_L), and is given as follows Equation (7) [39]:

$$RL = \frac{1}{1 + a_L C_o} \qquad (7)$$

where C_o is the initial dye concentration (mg/L) and a_L is the Langmuir constant related to the energy of adsorption (L/mg). The calculated R_L values indicate the type of adsorption: unfavourable process ($R_L > 1$), linear ($R_L = 1$), favourable (R_L between 0 and 1), or irreversible ($R_L = 0$) [40].

The Temkin formula determines that the decrease of adsorption heat with coverage is linear because of some adsorbate/adsorbent interactions. The adsorption is characterized by uniform distribution of bond energies, up to a maximum bond energy [41]. The linear form of Temkin isotherm's equation can be expressed as Equation (8), as follows:

$$q_e = \frac{RT}{b_T} \ln a_T + \frac{RT}{b_T} \ln C_e \qquad (8)$$

where T is the absolute temperature (K); R is the universal gas constant (8.314 J/mol K); a_T is the constant of Temkin isotherm (L/g); b_T is the Temkin constant related to the heat of adsorption (J/mol); and q_e and C_e are the equilibrium concentration of DB78 on eggshell (mg/g) and in the solution (mg/L), respectively. The Temkin constants a_T and b_T values can be calculated from the slope and intercept of straight line plot of q_e versus $\ln C_e$.

2.8. Thermodynamic Study

The thermodynamic analysis is needed to conclude whether the adsorption process of the dye onto eggshell is exothermic or endothermic. In order to gain further insight related to these experiments, it is essential to calculate the value of $\Delta H°$ (standard enthalpy change), $\Delta S°$ (standard entropy change), and $\Delta G°$ (Gibbs standard free energy change). The values for the different thermodynamic parameters can be calculated using the thermodynamic equilibrium coefficient obtained at different concentrations and temperatures. The $\Delta G°$ value is the fundamental parameter to elucidate the spontaneity of the adsorption process, and reactions occur spontaneously when the value of $\Delta G°$ is negative [34].

Considering the exchange adsorption, the equation employed to calculate the $K°$ value at different temperatures was as follows:

$$K° = K_p \times M_{adsorbate} \times 55.5 \tag{9}$$

where K_p is the equilibrium constant at time t (L/g), $M_{adsorbate}$ is the molecular weight of DB78 (g/mol), and 55.5 (mol/L) is the mole concentration of water [42].

Using the $K°$ values obtained from the previous equation, the Gibbs free energy was determined using the following equation:

$$\Delta G° = -RT \ln K° \tag{10}$$

To confirm the results for the Gibbs free energy, the Van't Hoff [12] equation was graphed ($\ln K°$ vs. $1/T$). Plotting $\ln K°$ versus $1/T$ gives a straight line with slope and intercept equal to $-\Delta H°/RT$ and $\Delta S°/R$, respectively, where R is the universal gas constant (8.314 J/mol K) and T is the absolute temperature in Kelvin (K). Using this representation, the Gibbs free energy was calculated again using the following equation:

$$\Delta G° = \Delta H° - T\Delta S° \tag{11}$$

2.9. Desorption and Regeneration of the Adsorbent

Desorption studies help to recover the adsorbate and adsorbent. Thus, the regeneration of the adsorbent may be important to reduce cost processes and to recover the pollutant extracted from the solution [37]. The viability of desorption and reuse of eggshell was evaluated using 0.5 M NaOH solution. First, 0.5 g of eggshell was stirred with 40 mL of dye (100 mg/L, pH 5), at 25 °C, 100 min of contact time, and 500 rpm for the adsorption phase. The adsorbent was dried and added into 40 mL of NaOH solution for desorption at a constant speed of 500 rpm for 100 min at 25 °C. The dye concentration in the solution was measured at 100 min, after the centrifugation of the samples at 3000 rpm for 5 min to separate the solid phase.

2.10. Degradation of DB78 by an Advanced Oxidation Process

The adsorption experiment was carried out using different dye concentrations. After 140 min of contact time and 300 mg/L of dye, the remaining dye in the solution was 144 mg/L. The experiments were performed in triplicate, which means that 144 mg/L was the average of the three repetitions. Then, 18 mL of the unadsorbed dye solution was mixed with 2 mL of a H_2O_2 solution that rendered a final H_2O_2 concentration of 840 mg/L. This rendered an H_2O_2/DB78 ratio of 200 on molar basis, which is enough to avoid making H_2O_2 the limiting reagent of the photochemical reaction. The mixture was placed in a pulsed light system (XeMaticA-Basic-1L, Steribeam, Germany) operated at 2.5 kV that supplied 2.14 J/cm^2 per pulse of a broad-spectrum light that included UV light, the spectrum of which has been previously reported [43]. Treatments were prolonged using multiple pulses up to 120 and carried out in duplicate. Samples were withdrawn every five pulses to measure absorbance. Data were normalized and adjusted to pseudo-first order kinetics (Equation (12)) to calculate the degradation rate.

$$\ln \frac{C}{C_0} = -kF \tag{12}$$

where C is the concentration at fluence F (J/cm^2), C_0 is the initial concentration, and k is the pseudo-first order rate constant (cm^2/J). Data were processed using Excel 2010 (Microsoft, Redmond, WA, USA).

3. Results and Discussion

3.1. Effect of Eggshell Dosage

Different amounts of eggshell (0.25, 0.5, 1.0, 1.5, and 2.0 g) were added into DB78 solutions to analyze the effect of adsorbent dosage on adsorption. The effect was investigated at an initial concentration of 25 mg/L, pH 5, 120 min of contact time, constant stir (500 rpm), and room temperature (25 °C). The results obtained from this study are shown in the plot of dye removal (%) versus adsorbent dosage (g), as can be seen from Figure 1.

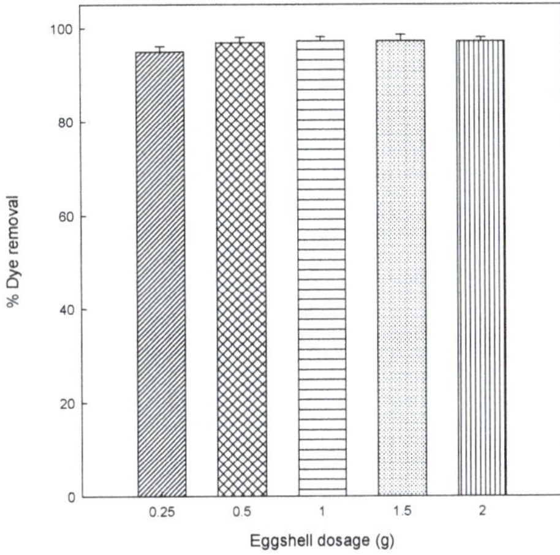

Figure 1. Effect of eggshell amount on the adsorption of Direct Blue 78 (DB78).

The increase of the adsorbent led to a slight increase for dye removed until a maximum adsorption. This reveals that the removal efficiency of dye increased with increasing amounts of eggshell. At a dosage of 0.5 g, the removal of dye was similar to higher doses (1–2 g), therefore, in order to optimize the adsorption process, and so to use the minimum amount of adsorbent, 0.5 g of eggshell was used in order to study the kinetic and equilibrium experiments.

3.2. Effect of Initial pH Solution

The pH of dye solution is considered an essential factor in any adsorption process that affects the adsorption capacity and the adsorption mechanism of the eggshell. With the objective to investigate the effect of the initial dye pH in the adsorption of DB78 dye on the eggshell surface, the equilibrium was studied at different pH values. The effect of pH in the adsorption of DB78 dye was evaluated in the range of pH from 3 to 11. The pH was adjusted by adding NaOH or HCl 0.1 M. The pH experiments were conducted using 2 g eggshell, at a concentration of 25 mg/L, 90 min of contact time, constant stir (500 rpm), and room temperature (25 °C), and the results are shown in Figure 2. The result indicates that the adsorption capacity and removal efficiency of eggshell depend on the pH.

As shown in Figure 2, the removal efficiency of dye increased at acidic pHs (3 and 5). From this pH, the adsorption capability of the eggshell decrease from 94.5% to 63.4% at pH 11. Figure 2 shows that the optimum pH required to obtain the maximum adsorption of DB78 onto eggshell is 5. The isoelectric point for the eggshell is 5.5 [44], which is close to our optimum pH (5.0). At this pH, the adsorbent is positively charged and interacts effectively with the negative charges in DB78. This trend was similar in the removal of Direct Red 80 and Acid Blue 25 dyes from aqueous solutions on eggshell membrane, in which the adsorption capacities of both anionic dyes onto eggshell membrane increased with acidic pHs [9] and, in another study, Chojnacka carried out the biosorption of Cr (III) ions from aqueous solutions by eggshells at pH 5 [45].

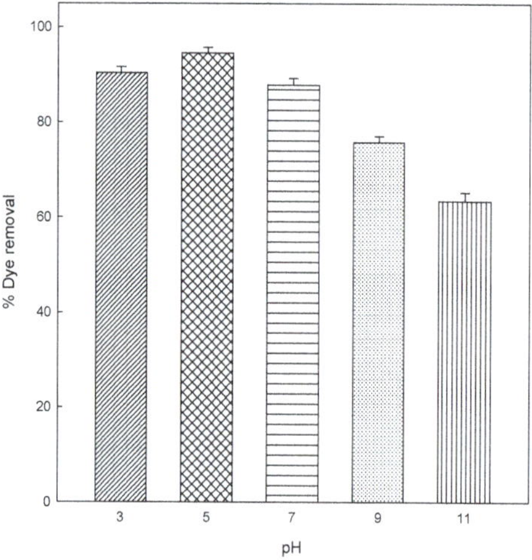

Figure 2. Effect of pH on the adsorption of DB78 by eggshell at different pH values.

The principal component of the shell is $CaCO_3$ in the form of the mineral calcite. In water, the carbonate species derived from calcite are H_2CO_3, HCO_3^-, and CO_3^{2-}, and a sufficient amount of carbonates is solubilized from the shells to buffer the mixtures to a low alkaline pH (from 7.5 to 8) after reaching the equilibrium. Therefore, at various pH, the electrostatic attraction as well as the structure of dye molecules and the eggshell could play very important roles in dye adsorption on this adsorbent. At pH 5, a significantly high electrostatic attraction exists between the positively charged surface of the adsorbent, owing to the ionization of functional groups of adsorbent and negatively charged dye. As the pH of the system increases, the number of negatively charged sites is increased. A negatively charged site on the adsorbent does not favor the adsorption of this dye owing to the electrostatic repulsion [46,47].

3.3. Effect of Contact Time

The following step in the adsorption experiments is to elucidate the effect of contact time using different concentrations of dye (from 25 to 300 mg/L). All the experiments were carried out at pH 5.0, constant stir (500 rpm), and room temperature (25 °C), with a fixed amount of adsorbent (0.5 g), and the results obtained are presented in Figure 3.

As can be observed in Figure 3, the adsorption capacity increased in each concentration until the equilibrium was achieved. In the equilibrium, the amount of adsorbed dye inside the adsorbent and the amount of dye desorbed were in a dynamic equilibrium. The time required for the adsorption of the dye onto the adsorbent to attain the equilibrium state is called the equilibrium time and the amount of dye removed by the adsorbent at that time indicates the maximum adsorption capacity of the adsorbent under these conditions [48].

Different adsorption stages can be differentiated in the range of concentrations analyzed (25–300 mg/L). At a low concentration (25 mg/L), the equilibrium was reached after 40 min of contact. The adsorption was fast at the concentration of 100 mg/L, however, it was slower than at low concentrations, reaching the equilibrium after 80 min of contact time. At high concentrations of dye (>100 mg/L), the curves did not present the asymptotic form. Hence, the equilibrium time increased with increasing concentrations of DB78 (from 150 to 300 mg/L).

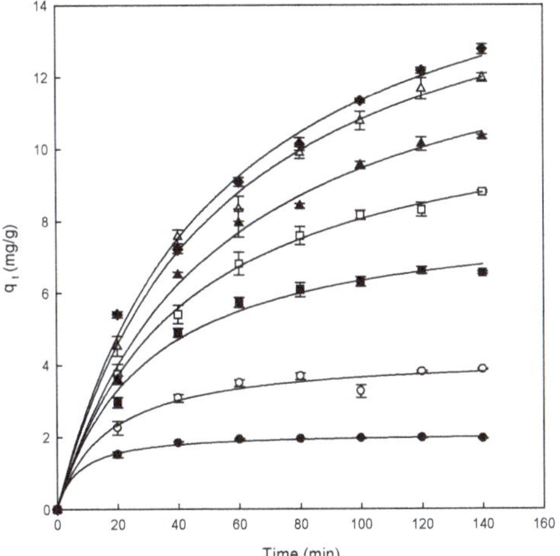

Figure 3. Effect of contact time between eggshell at different concentrations of Direct Blue 78 of 25 mg/L (●), 50 mg/L (○), 100 mg/L (■), 150 mg/L (□), 200 mg/L (▲), 250 mg/L (Δ), and 300 mg/L (♦).

3.4. Adsorption Kinetics

The adsorption surface, mass transfer, or intraparticle diffusion are different mechanisms involved in the adsorption process and, to study them, three kinetic models were employed to test the experimental data obtained in the adsorption of DB78 on eggshell. The determination coefficient values (R^2) are essential to decide the best adjustment to the experimental data to the different models tested. The results for the adjustment to the pseudo-first-order, pseudo-second-order, and the intraparticle diffusion kinetic models are shown in Figure 4, and the main parameters for each model are presented in Table 1.

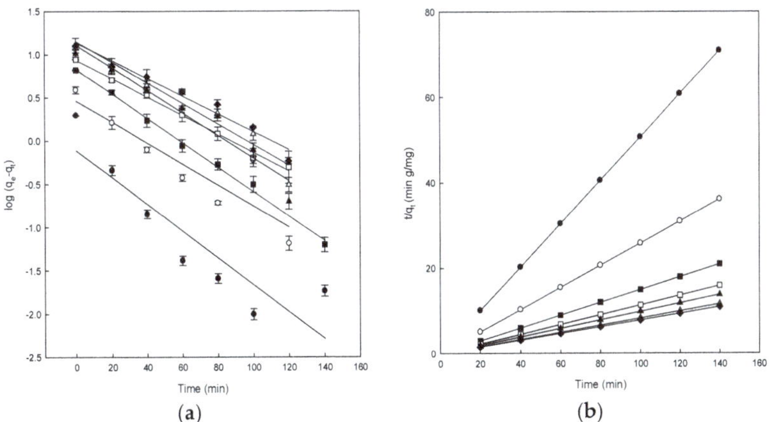

Figure 4. *Cont.*

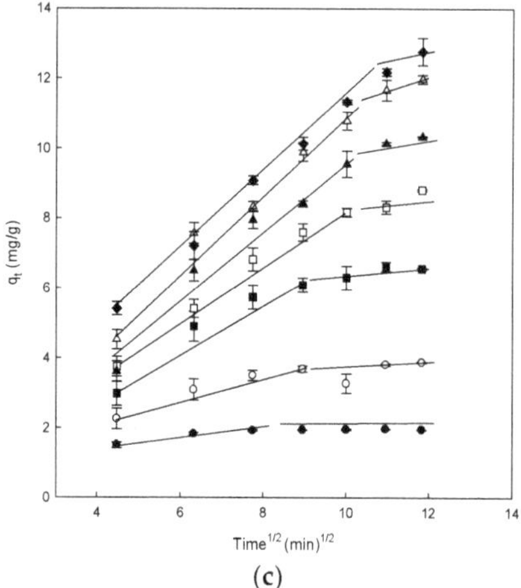

(c)

Figure 4. (**a**) Pseudo-first-order model plots, (**b**) pseudo-second-order model plots, and (**c**) intraparticle diffusion model plots for the Direct Blue 78 adsorption onto eggshell at different concentrations of dye of 25 mg/L (●), 50 mg/L (○), 100 mg/L (■), 150 mg/L (□), 200 mg/L (▲), 250 mg/L (Δ), and 300 mg/L (◆).

Table 1. Kinetics parameters of the pseudo-first-order, pseudo-second-order, and intraparticle diffusion models for the adsorption of Direct Blue 78 onto eggshell.

PFOM [1]	Eggshell			
C_0 (mg/L)	q_{eexp} (mg/g)	q_{ecal} (mg/g)	k_1 (min^{-1})	R^2
25	1.969	0.773	0.036	0.813
50	3.875	2.905	0.028	0.800
100	6.618	6.742	0.032	0.995
150	8.804	8.652	0.025	0.995
200	10.329	12.838	0.030	0.938
250	11.979	14.125	0.027	0.938
300	12.766	13.658	0.024	0.968
PSOM [2]				
C_0 (mg/L)	q_{eexp} (mg/g)	q_{ecal} (mg/g)	k_2 (g/mg min)	R^2
25	1.969	1.969	0.257	1
50	3.875	3.874	0.066	1
100	6.618	6.618	0.022	1
150	8.804	8.803	0.0129	1
200	10.329	10.331	0.0093	1
250	11.979	11.979	0.0069	1
300	12.766	12.771	0.0061	1
IDM [3]				
C_0 (mg/L)	q_{eexp} (mg/g)	(C) (mg/g)	k_i (mg/g min$^{1/2}$)	R^2
25	–	1.414	0.053	0.704
50	–	1.722	0.190	0.778
100	–	1.618	0.462	0.869
150	–	1.111	0.682	0.961
200	–	0.424	0.888	0.951
250	–	0.651	0.999	0.974
300	–	0.896	1.026	0.995

[1] PFOM: pseudo-first-order model; [2] PSOM: pseudo-second-order model; [3] IDM: intraparticle diffusion model.

The linearity of the Lagergren model ($\log(q_e - q_t)$ versus t) was graphed for 140 min of contact with the eggshell (Figure 4a). The R^2 values ranged from 0.800 to 0.995 (Table 1). Calculated values of q_e were compared with the experimental data, and although some R^2 values were relatively high, the

q_e values calculated were not suitable. The obtained R^2 and q_{ecal} values indicated that the adsorption of dye onto eggshell did not follow the pseudo-first-order kinetics, even though some values were relatively high; consequently, this equation cannot be used to analyze the experimental results. Because of the obtained results, it was appropriate to fit the experimental data to the pseudo-second-order model. Figure 4b shows the results obtained after applying the Ho and McKay model. The plot of t/q_t versus t produced straight lines for the entire measurement range.

The theorical q_e values were identical to the experimental q_e values obtained using the pseudo-second-order, as compared with those of the pseudo-first-order kinetic, indicating that DB78 adsorption onto eggshell followed the pseudo-second-order kinetic model. As can be observed from Table 1, the R^2 values of the pseudo-second-order kinetic model are higher than those of pseudo-first-order, the value was 1 in all cases analyzed. These results suggested that chemical adsorption was the rate-limiting step that controls this adsorption process. Chemisorption occurs when strong interactions, including hydrogen bonding and covalent and ionic bond formation, occur between the adsorbate and the solid surface. The endpoint for chemisorption is when all the active sites on the solid surface are occupied by chemisorbed molecules. Ehrampoush et al. observed similar kinetics in the adsorption of Reactive Red 123 dye onto eggshell [49], or in the adsorption of Acid Orange 51 onto the ground eggshell powder [16].

The adsorption is a process that follows many steps; firstly, it implicates a transport of dye molecules from the solution to the adsorbent surface, and then a diffusion to the interior of the eggshell could take place [32]. In order to understand the adsorption of DB78 dye onto eggshell, the kinetic of the adsorption process was analyzed using the intraparticle diffusion model, in order to determine if the intraparticle diffusion is the rate-limiting step in the adsorption. This effect was studied by plotting the amount of DB78 dye adsorbed versus the square root of time (Figure 4c).

Figure 4c shows the plot of q_t versus $t^{1/2}$ for the intraparticle diffusion of DB78 for the eggshell and different concentrations of dye. Two different straight lines can be distinguished in Figure 4c for the range of concentrations analyzed, indicating that two or more forces are influencing the adsorption process; in this case, chemisorption and intraparticle diffusion played essential roles in the adsorption of DB78 onto eggshell (Figure 4c).

The intraparticle diffusion constant (k_i) values are presented in Table 1. The values of k_i and C were calculated from the slope and intercept of plots of q_t versus $t^{1/2}$. These k_i values increased with increasing dye concentrations. The R^2 values were very different depending on the concentration, ranging from 0.704 to 0.995. The intraparticle diffusion model was not the rate-controlling step because the results did not pass through the origin. When this plot gives rise to a straight line, the adsorption process is controlled by intra-particle diffusion only. However, if the data present multi-linear plots, then two or more steps influence the adsorption process, as shown in Figure 4. Similar results were also reported for the adsorption of Acid Red 14 and Acid Blue 92 onto the microporous and mesoporous eggshell membrane [37].

Conversely, the intercept of each curve is proportional to the boundary layer thickness; a higher intercept indicates a higher effect. This value decreased with increasing dye concentrations for the eggshell; therefore, the intraparticle diffusion model was not the sole rate-controlling step for eggshell, confirming our previous suggestions [32].

3.5. Adsorption Equilibrium

The adsorption process of DB78 dye by eggshell adsorbent can be analyzed by fitting the experimental data of adsorption equilibrium to different isotherm models to find the most suitable isotherm to describe the adsorption process. Three well-known models were applied in the present study: Freundlich, Langmuir and Temkin isotherms. The constant isotherm parameters obtained from linear regression describe the equilibrium characteristics of adsorption, and are presented in Table 2. The plot of $\ln q_e$ versus $\ln C_e$ gave a straight line over the entire concentration range studied in the representation of the Freundlich isotherm, as may be observed in Figure 5a.

Table 2. Adsorption isotherm constants obtained for eggshell adsorbent.

Isotherm	Parameter	Eggshell
Freundlich	K_F (mg/g) (L/mg)$^{1/n}$	3.02
	n_F	3.40
	R^2	0.991
Langmuir	q_{max} (mg/g)	13.00
	K_L (L/g)	1.20
	a_L (L/mg)	0.093
	R^2	0.975
	R_L	0.281–0.032
Temkin	a_T (L/g)	6.73
	b_T (J/mol)	1.41
	R^2	0.951

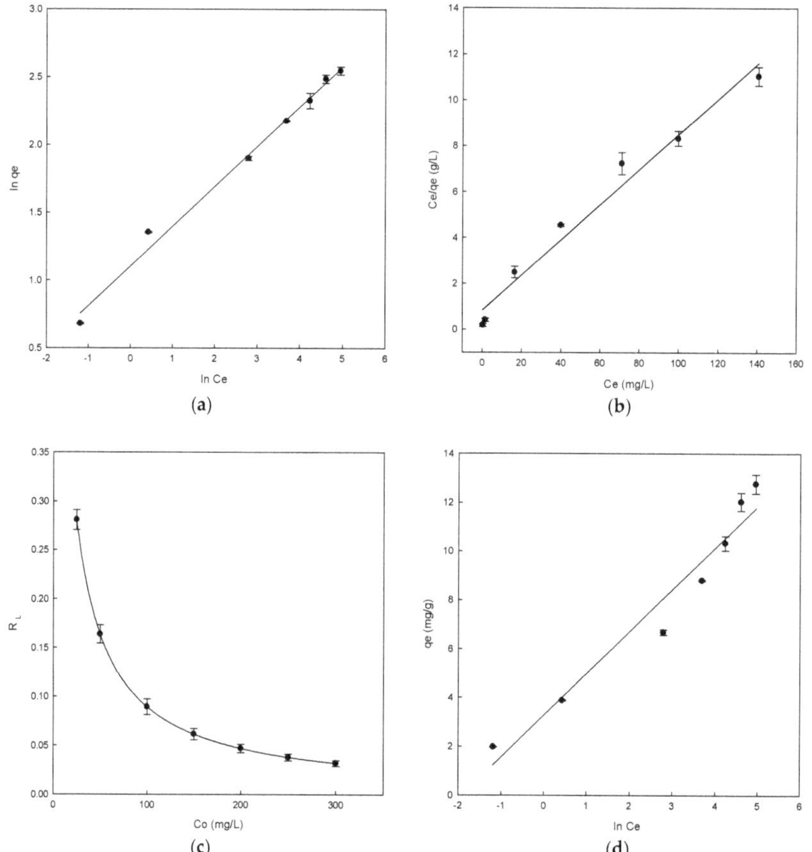

Figure 5. Adsorption isotherms for Direct Blue 78 by eggshell (**a**) Freundlich isotherm, (**b**) Langmuir isotherm, (**c**) separation factor, and (**d**) Temkin isotherm.

Thus, the straight line obtained was used to calculate the parameters K_F, n_F, and R^2. In this case, the K_F was 3.02 and n_F was 3.4 (Table 2). The process is favourable when the n_F value was found in the range between 1 and 10, which was confirmed for eggshell. A good linear determination coefficient ($R^2 = 0.991$) (Table 2) was obtained. The Freundlich model was the most suitable to describe the adsorption process owing to the high determination coefficient obtained. The linear form of the Langmuir isotherm was obtained by plotting C_e/q_e versus C_e, giving a straight line (Figure 5b). The Langmuir isotherm constants a_L/K_L and $1/K_L$ were determined from the slope and intercept of plot (C_e/q_e vs. C_e), respectively. K_L/a_L is q_{max} parameter, which is the maximum adsorption capacity of the adsorbent (mg/g) (Figure 5b). The parameters and constants obtained for Langmuir can be observed in Table 2. The maximum adsorption capacity of eggshell was 13 mg/g. Moreover, the value of R^2 was 0.975, but this value was lower than the Freundlich isotherm value.

This q_{max} value was similar to that of other adsorbents, for example, using chitosan in the adsorption of DB78 [15]; in the removal of various dyes using cellulose by-products, banana, and orange peels [50]; or in the adsorption of methylene blue dye using raw olive stone [20].

According to the results obtained, the Freundlich model fitted better than the Langmuir model. This may demonstrate the presence of heterogeneous adsorption sites on the eggshell surfaces. Similar results were analyzed for the adsorption of Reactive Red 123 dye using eggshell, in which the Freundlich and Langmuir isotherm models provided excellent fit with the highest R^2 value [49]. In another study, Pramanpol and Nitayapat carried out the adsorption of Reactive Yellow 205 dye using various components of eggshells and found a good concordance with the Freundlich model [51].

In the Langmuir isotherm, the value of R_L determines if the adsorption process is favourable or unfavourable. A value of R_L in the range between 0 and 1 indicates that the process is favourable. The values obtained for eggshell ranged from 0.281 to 0.032, confirming the adsorption process was favourable (Figure 5c). At low concentrations, the highest R_L values are also observed, indicating that the adsorption is more favourable at those concentrations.

The plot of q_e versus $\ln C_e$ shows the representation of the Temkin isotherm (Figure 5d). The Temkin constants b_T and a_T were determined from the slope and intercept, respectively. The b_T value obtained was 1.41 and the a_T value was 6.73, as shown in Table 2. A positive value of b_T indicates that physical and chemical forces were involved in the adsorption process. The value of R^2 was 0.951. The obtained result is lower than the Freundlich and Langmuir determination coefficients, with the Freundlich model being the best isotherm to explain the experimental results.

3.6. Thermodynamic Study

In order to study the effect of temperature on the adsorption of DB78 by eggshell, the experiments were conducted at three temperatures (Figure 6) at a concentration of 50 mg/L, pH 5, 80 min of contact time, and constant stir (500 rpm). The values of the thermodynamic parameters obtained at different temperatures are presented in Table 3.

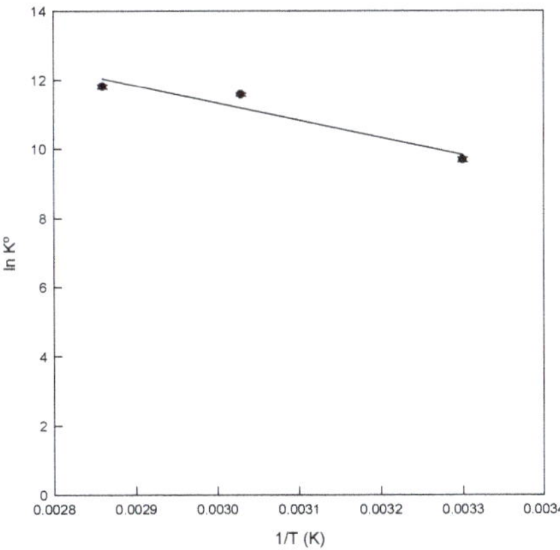

Figure 6. Van't Hoff plot for the adsorption of Direct Blue 78 onto eggshell at different temperatures.

Table 3. Thermodynamic parameters for the adsorption of Direct Blue 78 onto eggshell at different temperatures.

	T (K)	ΔG° (J/mol)	ΔH° (J/mol)	ΔS° (J/mol)
Eggshell	302	−24,417.19	41,690.71	219.24
	328	−30,886.69	-	-
	348	−34,376.63	-	-

The standard free energy (ΔG°) of the adsorption of DB78 onto eggshell was −24,417.19 J/mol, −30,886.69 J/mol, and −34376.63 J/mol at temperatures of 29, 55, and 75 °C, respectively. The exergonic values obtained of ΔG° indicated that the process is spontaneous at the three temperatures tested, confirming the viability of the process. The enthalpy change (ΔH°) was 41,690.71 J/mol. The positive value of ΔH° indicated the endothermic nature of the adsorption process. Table 3 shows that the value of ΔG° decreased with the increasing temperatures. The value decreased from −24,417.19 to −34,376.63 J/mol, which indicates a clear trend in the process. The adsorption process was favoured at high temperatures.

As stated before, the thermodynamic analysis revealed that, at 29 °C, there was not an increase in the adsorption abilities of eggshell, however, the best conditions to entrap the dye were achieved at 75 °C. At this temperature, the ability of the adsorbent to adsorb more dye molecules increased, as can be observed in Figure 7.

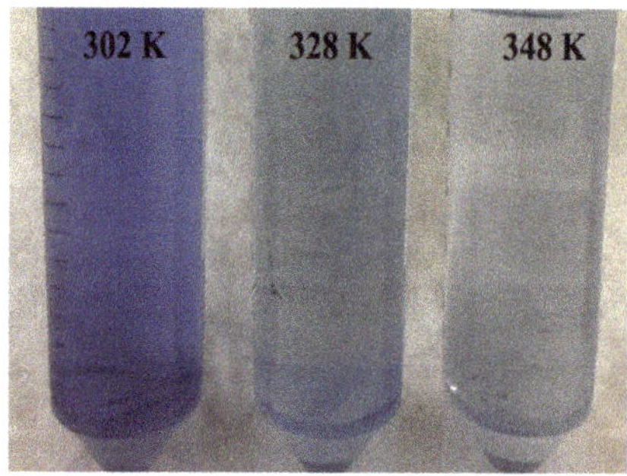

Figure 7. Effect of temperature for the adsorption of Direct Blue 78 onto eggshell at a concentration of 50 mg/L at different temperatures.

3.7. Desorption and Regeneration of the Adsorbent

Prior to the desorption measurements, it was mandatory to carry out adsorption experiments. After this first adsorption cycle, the adsorbent was put in an alkaline medium (pH 12). In the first desorption cycle, 21% of dye was released from the eggshell; this could explain the electrostatic repulsion between the anionic dye and the negatively charged surface on the adsorbent. The ability of the eggshell to desorb more dye molecules after four consecutive cycles could be attributed to the distribution of charges on the adsorbent surface. After exposing the adsorbent to the same pH solution four times, the adsorbent releases the dye more easily. Arami et al. observed similar results in the adsorption of two different dyes onto the microporous and mesoporous eggshell membrane [37].

This adsorption/desorption cycle was repeated four times (Figure 8); according to the results observed, the adsorption abilities of the eggshell decreased with the increasing the number of cycles. This trend was the opposite for the desorption experiments. The results confirm that the eggshell can be reused in different cycles of adsorption/desorption.

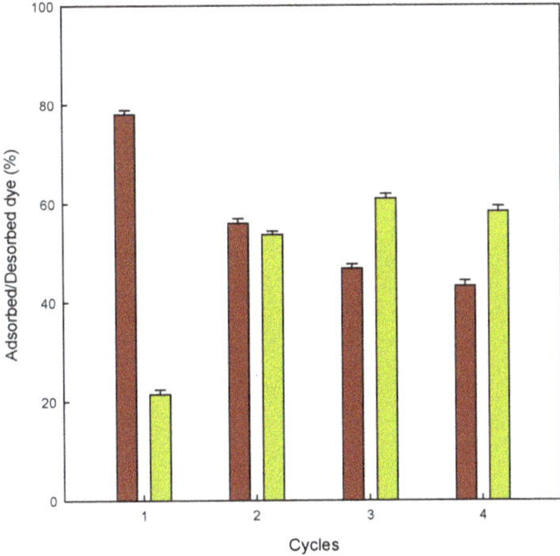

Figure 8. Adsorbed/desorbed dye (%) onto eggshell for four consecutive adsorption (■)/desorption (■) cycles.

3.8. Degradation of DB78 by an H_2O_2/Pulsed Light AOP

The degradation of DB78 by the H_2O_2/pulsed light AOP can be observed in Figure 9. The process was able to degrade the dye by more than 95%, which can be considered a significant reduction in the contamination potential of this dye if released to the environment. The respective pseudo-first order degradation rate was 0.012 cm^2/J (R^2 = 0.9967). Prolonging the treatment would not be efficient as data extrapolation allows predicting 99% degradation after supplying an energy of 383 J/cm^2, which means expending 49% more energy to increase the degradation by only 4%.

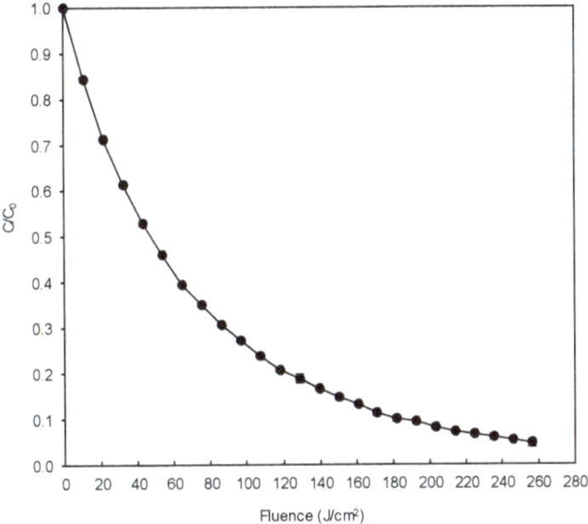

Figure 9. Degradation of DB78 by the H_2O_2/pulsed light advanced oxidation process in model wastewater.

4. Conclusions

The present study showed that eggshell can be successfully utilized as a biosorbent for the removal of DB78 dye from water solutions. It was found that the adsorption of DB78 onto eggshell followed pseudo-second-order kinetics. The Freundlich isotherm was the best model to describe adsorption. The maximum adsorption of DB78 onto eggshell was obtained at pH 5 and 0.5 g of adsorbent dosage. According to the Langmuir isotherm, the maximum adsorption capacity of DB78 onto eggshell was 13 mg/g. Taking into account the results obtained for the n_F and R_L parameters, the adsorption is considered as favorable. The analysis of thermodynamic parameters revealed that the adsorption process is endothermic and spontaneous at the three temperatures analyzed. Desorption studies were conducted and the results showed that the eggshell was reusable in different adsorption/desorption cycles. Complementing the adsorption process by an H_2O_2/pulsed light advanced oxidation process allows further decreasing the release of pollutant dyes to the environment, proving that the combination of both techniques can be used successfully in the removal of dyes from wastewater at higher concentrations of dye.

Author Contributions: The authors who sign the following manuscript have made significant contributions, allowing achieving the set objectives: A.M.-S. carried out the adsorption experiments; J.A.P. wrote the manuscript and interpreted the kinetic and isotherm results obtained; M.I.R.-L. graphed the results; V.M.G.-L. carried out the photolysis of remaining dye by using pulsed light; E.N.-D., and J.A.G. were responsible for the conception, design, assessment, and revision of the work. All authors have read and agreed to the published version of the manuscript.

Funding: This research received no external funding.

Conflicts of Interest: The authors declare no conflict of interest.

References

1. Vakili, M.; Rafatullah, M.; Salamatinia, B.; Abdullah, A.Z.; Ibrahim, M.H.; Tan, K.B.; Gholami, Z.; Amouzgar, P. Application of chitosan and its derivatives as adsorbents for dye removal from water and wastewater: A review. *Carbohydr. Polym.* **2014**, *113*, 115–130. [CrossRef] [PubMed]

2. Nakkeeran, E.; Varjani, S.J.; Dixit, V.; Kalaiselvi, A. Synthesis, characterization and application of zinc oxide nanocomposite for dye removal from textile industrial wastewater. *Indian J. Exp. Biol.* **2018**, *56*, 498–503.

3. Li, X.; Jin, X.; Zhao, N.; Angelidaki, I.; Zhang, Y. Novel bio-electro-Fenton technology for azo dye wastewater treatment using microbial reverse-electrodialysis electrolysis cell. *Bioresour. Technol.* **2017**, *228*, 322–329. [CrossRef] [PubMed]

4. Homaeigohar, S. The nanosized dye adsorbents for water treatment. *Nanomaterials* **2020**, *10*, 295. [CrossRef]

5. Homaeigohar, S.; Botcha, N.K.; Zarie, E.; Elbahri, M. Ups and downs of water photodecolorization by nanocomposite polymer nanofibers. *Nanomaterials* **2019**, *9*, 250. [CrossRef]

6. Homaeigohar, S.; Zillohu, A.U.; Abdelaziz, R.; Hedayati, M.K.; Elbahri, M. A novel nanohybrid nanofibrous adsorbent for water purification from dye pollutants. *Materials* **2016**, *9*, 848. [CrossRef]

7. Moussavi, G.; Mahmoudi, M. Removal of azo and anthraquinone reactive dyes from industrial wastewaters using MgO nanoparticles. *J. Hazard. Mater.* **2009**, *168*, 806–812. [CrossRef]

8. Mahmoud, H.R.; Ibrahim, S.M.; El-Molla, S.A. Textile dye removal from aqueous solutions using cheap MgO nanomaterials: Adsorption kinetics, isotherm studies and thermodynamics. *Adv. Powder Technol.* **2016**, *27*, 223–231. [CrossRef]

9. Arami, M.; Limaee, N.Y.; Mahmoodi, N.M. Investigation on the adsorption capability of egg shell membrane towards model textile dyes. *Chemosphere* **2006**, *65*, 1999–2008. [CrossRef]

10. Ngah, W.W.; Teong, L.; Hanafiah, M.A.K.M. Adsorption of dyes and heavy metal ions by chitosan composites: A review. *Carbohydr. Polym.* **2011**, *83*, 1446–1456. [CrossRef]

11. Carvalho, J.; Araujo, J.; Castro, F. Alternative low-cost adsorbent for water and wastewater decontamination derived from eggshell waste: An overview. *Waste Biomass Valoriz.* **2011**, *2*, 157–167. [CrossRef]

12. Yagub, M.T.; Sen, T.K.; Ang, M. Removal of cationic dye methylene blue (MB) from aqueous solution by ground raw and base modified pine cone powder. *Environ. Earth Sci.* **2014**, *71*, 1507–1519. [CrossRef]

13. Burakov, A.E.; Galunin, E.V.; Burakova, I.V.; Kucherova, A.E.; Agarwal, S.; Tkachev, A.G.; Gupta, V.K. Adsorption of heavy metals on conventional and nanostructured materials for wastewater treatment purposes: A review. *Ecotoxicol. Environ. Saf.* **2018**, *148*, 702–712. [CrossRef] [PubMed]

14. Noreen, S.; Bhatti, H.N. Fitting of equilibrium and kinetic data for the removal of Novacron Orange P-2R by sugarcane bagasse. *J. Ind. Eng. Chem.* **2014**, *20*, 1684–1692. [CrossRef]

15. Murcia-Salvador, A.; Pellicer, J.A.; Fortea, M.I.; Gómez-López, V.M.; Rodríguez-López, M.I.; Núñez-Delicado, E.; Gabaldón, J.A. Adsorption of Direct Blue 78 using chitosan and cyclodextrins as adsorbents. *Polymers* **2019**, *11*, 1003. [CrossRef] [PubMed]

16. Tsai, W.T.; Hsien, K.J.; Hsu, H.C.; Lin, C.M.; Lin, K.Y.; Chiu, C.H. Utilization of ground eggshell waste as an adsorbent for the removal of dyes from aqueous solution. *Bioresour. Technol.* **2008**, *99*, 1623–1629. [CrossRef] [PubMed]

17. De Oliveira, A.P.; Módenes, A.N.; Bragião, M.E.; Hinterholz, C.L.; Trigueros, D.E.; Isabella, G.D.O. Use of grape pomace as a biosorbent for the removal of the Brown KROM KGT dye. *Bioresour. Technol. Rep.* **2018**, *2*, 92–99. [CrossRef]

18. Kyzas, G.Z.; Lazaridis, N.K.; Mitropoulos, A.C. Removal of dyes from aqueous solutions with untreated coffee residues as potential low-cost adsorbents: Equilibrium, reuse and thermodynamic approach. *Chem. Eng. J.* **2012**, *189*, 148–159. [CrossRef]

19. El-Bindary, A.A.; El-Sonbati, A.Z.; Al-Sarawy, A.A.; Mohamed, K.S.; Farid, M.A. Adsorption and thermodynamic studies of hazardous azocoumarin dye from an aqueous solution onto low cost rice straw based carbons. *J. Mol. Liq.* **2014**, *199*, 71–78. [CrossRef]

20. Hazzaa, R.; Hussein, M. Adsorption of cationic dye from aqueous solution onto activated carbon prepared from olive stones. *Environ. Technol. Innov.* **2015**, *4*, 36–51. [CrossRef]

21. Van Thuan, T.; Quynh, B.T.P.; Nguyen, T.D.; Bach, L.G. Response surface methodology approach for optimization of Cu2+, Ni2+ and Pb2+ adsorption using KOH-activated carbon from banana peel. *Surf. Interfaces* **2017**, *6*, 209–217. [CrossRef]

22. Fernández-López, J.A.; Angosto, J.M.; Roca, M.J.; Miñarro, M.D. Taguchi design-based enhancement of heavy metals bioremoval by agroindustrial waste biomass from artichoke. *Sci. Total Environ.* **2019**, *653*, 55–63. [CrossRef] [PubMed]

23. Yu, J.X.; Zhu, J.; Feng, L.Y.; Chi, R.A. Simultaneous removal of cationic and anionic dyes by the mixed sorbent of magnetic and non-magnetic modified sugarcane bagasse. *J. Colloid Interface Sci.* **2015**, *451*, 153–160. [CrossRef] [PubMed]

24. Zhang, Y.J.; Ou, J.L.; Duan, Z.K.; Xing, Z.J.; Wang, Y. Adsorption of Cr (VI) on bamboo bark-based activated carbon in the absence and presence of humic acid. *Colloid Surf. A* **2015**, *481*, 108–116. [CrossRef]

25. Mittal, A.; Teotia, M.; Soni, R.; Mittal, J. Applications of egg shell and egg shell membrane as adsorbents: A review. *J. Mol. Liq.* **2016**, *223*, 376–387. [CrossRef]

26. Martínez-López, S.; Lucas-Abellán, C.; Serrano-Martínez, A.; Mercader-Ros, M.T.; Cuartero, N.; Navarro, P.; Pérez, S.; Gabaldón, J.A.; Gómez-López, V.M. Pulsed light for a cleaner dyeing industry: Azo dye degradation by an advanced oxidation process driven by pulsed light. *J. Clean. Prod.* **2019**, *217*, 757–766. [CrossRef]

27. Daengprok, W.; Garnjanagoonchorn, W.; Mine, Y. Fermented pork sausage fortified with commercial or hen eggshell calcium lactate. *Meat Sci.* **2002**, *62*, 199–204. [CrossRef]

28. Balaz, M. Eggshell membrane biomaterial as a platform for applications in materials science. *Acta Biomater.* **2014**, *10*, 3827–3843. [CrossRef]

29. Livani, M.J.; Ghorbani, M. Fabrication of NiFe$_2$O$_4$ magnetic nanoparticles loaded on activated carbon as novel nanoadsorbent for Direct Red 31 and Direct Blue 78 adsorption. *Environ. Technol.* **2018**, *39*, 2977–2993. [CrossRef]

30. Ho, Y. Citation review of Lagergren kinetic rate equation on adsorption reactions. *Scientometrics* **2004**, *59*, 171–177. [CrossRef]

31. Ho, Y.; McKay, G. A comparison of chemisorption kinetic models applied to pollutant removal on various sorbents. *Process. Saf. Environ. Prot.* **1998**, *76*, 332–340. [CrossRef]

32. Crini, G.; Peindy, H.N.; Gimbert, F.; Robert, C. Removal of CI Basic Green 4 (Malachite Green) from aqueous solutions by adsorption using cyclodextrin-based adsorbent: Kinetic and equilibrium studies. *Sep. Purif. Technol.* **2007**, *53*, 97–110. [CrossRef]

33. Weber, W.J.; Morris, J.C. Kinetics of adsorption on carbon from solution. *J. Sanit. Eng. Div.* **1963**, *89*, 31–60.

34. Oladoja, N.; Aboluwoye, C.; Oladimeji, Y. Kinetics and isotherm studies on methylene blue adsorption onto ground palm kernel coat. *Turk. J. Eng. Env. Sci.* **2009**, *32*, 303–312.

35. Crini, G.; Badot, P.M. Application of chitosan, a natural aminopolysaccharide, for dye removal from aqueous solutions by adsorption processes using batch studies: A review of recent literature. *Prog. Polym. Sci.* **2008**, *33*, 399–447. [CrossRef]

36. Ho, Y.; Porter, J.; McKay, G. Equilibrium isotherm studies for the sorption of divalent metal ions onto peat: Copper, nickel and lead single component systems. *Water Air Soil Pollut.* **2002**, *141*, 1–33. [CrossRef]

37. Arami, M.; Limaee, N.Y.; Mahmoodi, N.M. Evaluation of the adsorption kinetics and equilibrium for the potential removal of acid dyes using a biosorbent. *Chem. Eng. J.* **2008**, *139*, 2–10. [CrossRef]

38. Dada, A.; Olalekan, A.; Olatunya, A.; Dada, O. Langmuir, Freundlich, Temkin and Dubinin–Radushkevich isotherms studies of equilibrium sorption of Zn^{2+} unto phosphoric acid modified rice husk. *IOSR J. Appl. Chem.* **2012**, *3*, 38–45.

39. McKay, G. Adsorption of dyestuffs from aqueous solutions with activated carbon I: Equilibrium and batch contact-time studies. *J. Chem. Technol. Biotechnol.* **1982**, *32*, 759–772. [CrossRef]

40. Pellicer, J.A.; Rodríguez-López, M.I.; Fortea, M.I.; Hernández, J.A.G.; Lucas-Abellán, C.; Mercader-Ros, M.T.; Serrano-Martínez, A.; Núñez-Delicado, E.; Cosma, P.; Fini, P. Removing of Direct Red 83: 1 using α-and HP-α-CDs polymerized with epichlorohydrin: Kinetic and equilibrium studies. *Dyes Pigments* **2018**, *149*, 736–746. [CrossRef]

41. Temkin, M. Kinetics of ammonia synthesis on promoted iron catalysts. *Acta Physicochim. URS* **1940**, *12*, 327–356.

42. Zhou, X.; Zhou, X. The unit problem in the thermodynamic calculation of adsorption using the Langmuir equation. *Chem. Eng. Comm.* **2014**, *201*, 1459–1467. [CrossRef]

43. Cudemus, E.; Izquier, A.; Medina-Martínez, M.S.; Gómez-López, V.M. Effects of shading and growth phase on the microbial inactivation by pulsed light. *Czech J. Food Sci.* **2013**, *31*, 189–193. [CrossRef]

44. Wang, W.; Chen, B.; Huang, Y.; Cao, J. Evaluation of eggshell membrane-based bio-adsorbent for solid-phase extraction of linear alkylbenzene sulfonates coupled with high-performance liquid chromatography. *J. Chromatogr. A* **2010**, *1217*, 5659–5664. [CrossRef]

45. Chojnacka, K. Biosorption of Cr (III) ions by eggshells. *J. Hazard. Mat.* **2005**, *121*, 167–173. [CrossRef]

46. Lunge, S.; Thakre, D.; Kamble, S.; Labhsetwar, N.; Rayalu, S. Alumina supported carbon composite material with exceptionally high defluoridation property from eggshell waste. *J. Hazard. Mater.* **2012**, *237*, 161–169. [CrossRef] [PubMed]

47. Eletta, O.A.A.; Ajayi, O.A.; Ogunleye, O.O.; Akpan, I.C. Adsorption of cyanide from aqueous solution using calcinated eggshells: Equilibrium and optimisation studies. *J. Environ. Chem. Eng.* **2016**, *4*, 1367–1375. [CrossRef]

48. Pellicer, J.A.; Rodríguez-López, M.I.; Fortea, M.I.; Lucas-Abellán, C.; Mercader-Ros, M.T.; López-Miranda, S.; Gómez-López, V.M.; Semeraro, P.; Cosma, P.; Fini, P. Adsorption properties of β-and hydroxypropyl-β-cyclodextrins cross-linked with epichlorohydrin in aqueous solution. A sustainable recycling strategy in textile dyeing process. *Polymers* **2019**, *11*, 252. [CrossRef]

49. Ehrampoush, M.; Ghanizadeh, G.; Ghaneian, M. Equilibrium and kinetics study of reactive red 123 dye removal from aqueous solution by adsorption on eggshell. *J. Environ. Health Sci. Eng.* **2011**, *8*, 101–106.

50. Annadurai, G.; Juang, R.S.; Lee, D.J. Use of cellulose-based wastes for adsorption of dyes from aqueous solutions. *J. Hazard. Mater.* **2002**, *92*, 263–274. [CrossRef]

51. Pramanpol, N.; Nitayapat, N. Adsorption of reactive dye by eggshell and its membrane. *Kasetsart J.* **2006**, *40*, 192–197.

 *materials*

Article

Recycled Expanded Polystyrene as Lightweight Aggregate for Environmentally Sustainable Cement Conglomerates

Andrea Petrella *, Rosa Di Mundo and Michele Notarnicola

Dipartimento di Ingegneria Civile, Ambientale, Edile, del Territorio e di Chimica, Politecnico di Bari, Via E. Orabona, 4, 70125 Bari, Italy; rosa.dimundo@poliba.it (R.D.M.); michele.notarnicola@poliba.it (M.N.)

* Correspondence: andrea.petrella@poliba.it; Tel.: +39-(0)-8059-63275; Fax: +39-(0)-8059-63635

Received: 20 January 2020; Accepted: 20 February 2020; Published: 22 February 2020

Abstract: In the present work the rheological, thermo-mechanical, microstructural, and wetting characteristics of cement mortars with recycled expanded polystyrene (EPS) were analyzed. The samples were prepared after partial/total replacement of the conventional sand aggregate with EPS having different grain size and size distribution. Lightness and thermal insulation were relevant features for all the bare EPS composites, despite the mechanical strengths. Specifically, EPS based mortars were characterized by higher thermal insulation with respect to the sand reference due to the lower specific mass of the specimens mainly associated with the low density of the aggregates and also to the spaces at the EPS/cement paste interfaces. Interesting results in terms of low thermal conductivity and high mechanical resistances were obtained in the case of sand-EPS mixtures although characterized by only 50% in volume of the organic aggregate. Moreover, sand-based mortars showed hydrophilicity (low WCA) and high water penetration, whereas the presence of EPS in the cement composites led to a reduction of the absorption of water especially on the bulk of the composites. Specifically, mortars with EPS in the 2–4 mm and 4–6 mm bead size range showed the best results in term of hydrophobicity (high WCA) and no water penetration in the inner surface, due to low surface energy of the organic aggregate together with a good particle distribution. This was indicative of cohesion between the ligand and the polystyrene as observed in the microstructural detections. Such a property is likely to be correlated to the observed good workability of this type of mortar and to its low tendency to segregation compared to the other EPS containing specimens. These lightweight thermo-insulating composites can be considered environmentally sustainable materials because they are prepared with no pre-treated secondary raw materials and can be used for indoor applications.

Keywords: recycled expanded polystyrene; cement mortars; safe production; thermal insulation; mechanical resistance

---

1. Introduction

In recent years, the problems associated with waste management have become very relevant in the frame of a more sustainable model of development and consumption of new resources and energy [1–7]. The construction industry is one of the activities with the greatest consumption of raw materials together with large production of waste [8–14]. Specifically, the broad use of plastics in building/construction applications, especially expanded polystyrene (EPS), requires new and low environmental impact approaches for the optimization of the production processes and the reduction of by-products [15–18]. For this reason, recycling operations can be considered important tasks to increase the sustainability of a material which is converted into a new resource, the so-called secondary raw material. For the purpose, expanded polystyrene is a completely recyclable material, widely used because of cost-effectiveness, versatility and performance characteristics [18–21]. It is manufactured from styrene monomer using a process during which pentane gas is added to the polymer in order to induce expansion with the following production of spherical beads. EPS is a thermoplastic polymer widely used in many applications (buildings, packaging) due to the relevant features as thermal insulation, durability, lightness, strength, shock absorption, and processability which allow high performance and economic products to be obtained [22–27]. EPS is a closed cell material with low water absorption and high resistance to moisture, thus preserving shape, size, and structure after water saturation. EPS resins are widespread polymers in the building/construction field and in civil engineering, usually available in sheet form, shapes or large blocks and used for floor insulation, closed cavity walls, roofs, etc., but also employed in road foundations, pavement construction, impact sound insulation, drainage, modular construction elements, lightweight conglomerates (concretes, mortars), etc. [28–34].

In this work, lightweight cement mortars containing recycled expanded polystyrene (EPS) from the grinding of industrial scraps were prepared with partial or complete replacement of the standard sand aggregate in the mixture, with no addition of additives. A study on the rheological, thermo-mechanical, microstructural, and wetting properties of the samples was conducted. The effect of the aggregate size and size distribution was evaluated and a comparison with specimens based on conventional and/or normalized sand was carried-out.

The aim was to realize an environmental, sustainable material with low specific mass and thermal insulating properties and characterized by high technical performances in term of hydrophobicity, low water absorption [35–39], and with a low impact manufacturing process. On the contrary to that observed in conventional cement composites, characterized by porosity and hydrophilicity, hydrophobic composites usually show longer durability together with self-cleaning properties [40,41]. Cement structure protection follows standard protocols based on impregnation/coating of external layers by silane or siloxane moieties thus leaving a hydrophilic consolidated concrete composite [41,42]. The addition of polymers to the fresh mixture together with the application of hydrophobic coatings on the hardened artifacts has been shown to lead to a reduction of water penetration thus converting the reference building material to have a hydrophobic or over-hydrophobic nature [43,44]. In the present research, the conglomerate did not show any coating on the surface and the whole mass was modified, for this reason the side and the fracture surfaces were investigated.

These lightweight thermo-insulating composites can be considered environmentally sustainable materials for indoor non-structural artifacts because they are prepared with non-pre-treated secondary raw materials and with a cheap route since complex techniques of production are not required. These treatments and processes would, however, be more effective in the case of production on a larger scale.

2. Materials and Methods

2.1. Preparation of the Mortars

Cement mortars were prepared with CEM II A-LL 42.5 R (Buzzi Unicem (Casale Monferrato, Italy)) [45]. Normalized sand (~1700 g/dm^3, 0.08–2 mm) was purchased by Societè Nouvelle du Littoral (Leucate, France), whereas sieved sand was used as aggregate in three specific size fractions (1–2 mm, 2–4 mm and 4–6 mm) [46,47]. Recycled expanded polystyrene (EPS), resulting from grinding of industrial scraps, was employed in three specific size fractions (1–2 mm, 2–4 mm, and 4– 6 mm). The specimens were prepared with a 0.5 W/C ratio and 40 mm × 40 mm ×160 mm prisms were obtained for the flexural/compressive tests, while cylinders (diameter = 100 mm; height = 50 mm) were prepared for the thermal tests. In the case of the mechanical tests the samples were cured in water for 7, 28, 45, and 60 days, while in the case of the thermal tests the samples were cured in water for 28 days.

The reference was prepared with normalized sand [46] and named Normal. EPS was added into the conglomerate with a partial or complete replacement of the standard sand aggregate which was made on volume basis rather than on weight basis [48–50] due to the low specific mass of polystyrene. The samples (excluding the Normal) were prepared with a volume of aggregate of 500 cm^3. Tables 1 and 2 show the composition of the aggregate and of the relative mortars.

Total sand replacement was carried out with EPS grains in the 1–2 mm (30 g/dm^3), 2–4 mm (15 g/dm^3) and 4–6 mm (10 g/dm^3) bead size range and EPS2, EPS3, and EPS4 samples were obtained as reported in Tables 1 and 2. Another reference, named Sand, prepared with sand size in the range of 1–2 mm (50%), 2–4 mm (25%), and 4–6 mm (25%), was compared to the EPS specimens. A Sand-EPS sample was prepared after replacement of the 50% sand volume with 4–6 mm EPS grains (Sand/EPS).

Table 1. Composition of the aggregates in the composites.

Normal	Normalized Sand		
Sand	sand (1–2 mm) 25%	sand (2–4 mm) 25%	sand (4–6 mm) 50%
Sand-EPS	sand(1–2 mm) 25%	sand(2–4 mm) 25%	EPS (4–6 mm) 50%
EPS 2			EPS (4–6 mm) 100%
EPS 3		EPS (2–4 mm) 50%	EPS (4–6 mm) 50%
EPS 4	EPS (1–2 mm) 25%	EPS (2–4 mm) 25%	EPS (4–6 mm) 50%

Table 2. Composition of the mortars.

Sample	Cement (g)	Water (cm^3)	Sand Volume (cm^3)	EPS Volume (cm^3)	ρ (Kg/m^3)	Porosity %
Normal	450	225	810	0	2020	22
Sand	450	225	500	0	2090	20
Sand-EPS	450	225	250	250	1320	32
EPS 2	450	225	0	500	850	49
EPS 3	450	225	0	500	940	42
EPS 4	450	225	0	500	855	48

2.2. Rheological, Thermal and Mechanical Characterization

The flow-test allowed the evaluation of the rheological properties of the fresh conglomerates [51]. An ISOMET 2104, Applied Precision Ltd (Bratislava, Slovakia), was used to determine the thermal conductivity (λ) and the thermal diffusivity (α) of the specimens by production of a constant thermal flow by a heating probe applied on the sample surface. The temperature was recorded over time and λ and α were obtained after evaluation of the experimental temperature compared with the solution of the heat conduction equation [52]. The flexural and compression tests were carried out by the use of a MATEST device (Milan, Italy). Flexural tests were carried out on six prisms (40 mm × 40 mm × 160 mm) by applying a load with a 50 ± 10 N/s rate, while compression strengths were obtained on the resulting semi-prisms by applying a load with a 2400 ± 200 N/s rate [46].

2.3. Measurements of the Contact Angle and of the Water Absorption

In the present research, an investigation on the side surface and on the inner surface of the cement conglomerates was carried out by contact angle measurements. After deposition of at least fifteen drops (5 µL) of water onto the surface of each specimen, it was shown that the behavior of three representative points (point 1, 2, and 3) summarized the behavior of all the drops. A Premier series dyno-lite (Taiwan) portable microscope and background cold lighting was used to study the time evolution of the drop, with a rate ranging of 30 frames per second. In the case of non- static drop, determined by water absorption, the image sequences were analyzed by Image J software (version 1.8.0, National Institute of Health, Bethesda, MD, USA) in order to measure the variation of the contact angle and of the drop height.

2.4. SEM/EDX and Porosimetric Analyses

An electron microscope FESEM-EDX Carl Zeiss Sigma 300 VP (Carl Zeiss Microscopy GmbH (Jena, Germany)) was used to characterize the morphology and the chemical composition of the samples which were applied onto aluminum stubs and sputtered with gold (Sputter Quorum Q150 Quorum Technologies Ltd (East Sussex, UK)) before the test. In this respect, the composition of the normalized sand was: C (4%), O (52%), Si (35%), Ca (2%), the composition of the sieved sand was: C (10%), O (45%), Ca (45%), polystyrene composition was: C (30%), O (70%), the composition of the cement paste was: C (4.2%), O (40%), Si (7.6%), Ca (44%), Fe (1.5%), Al (2.5%). Ultrapyc 1200e Automatic Gas Pycnometer, Quantachrome Instruments (Boynton Beach, FL, USA) was used for the porosimetric measurements and helium was used to penetrate the material pores.

3. Results and Discussion

Flow data of the non-consolidated samples are reported in Figure 1 and were obtained after measuring the diameters of the mixture before and after the test [51]. The flow of a sample is represented by the percentage increase of the diameter over the base diameter.

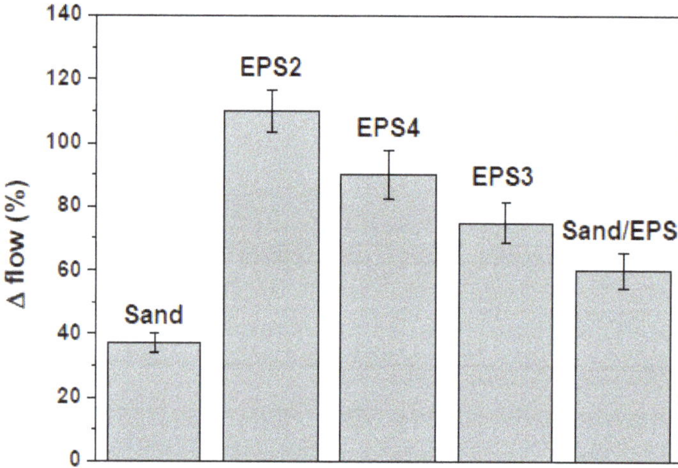

Figure 1. Flow-test results.

The Sand sample showed a higher flow (+35%) with respect to the Normal sample due to the absence of finer aggregates. The EPS specimens were more fluid than both references and specifically with respect to the normalized mortar (Normal). This behavior may be ascribed to the low surface energy, the low roughness (smooth surface), the hydrophobic features (synthetic organic polymer) and the low density of the EPS particles (10–30 g/dm3 with respect to 1700 g/dm^3 of sand) which may induce aggregate segregation in a cement conglomerate. The lower fluidity of EPS3 (+126%) with respect to EPS2 (+174%) and EPS4 (+150%) is likely due to a better compaction of the aggregates in the mixture (better distribution of the granules), while in the case of the Sand/EPS specimen, the presence of the inorganic aggregate contributed to a reduction of fluidity (Figure 1). In Figure 2 and Table 3 the flexural and compressive strengths of the samples are reported as a function of the specific mass. The Sand sample showed slightly higher mechanical strengths than the Normal sample due to the presence of larger size aggregates which contribute to the increase of the specific mass. The addition of EPS determined the formation of voids in the composite with a sensible reduction of the specific mass of the mortars (Table 2) which depends not only on the matrix and polymer characteristics (foaming structure of EPS), but also on the interface properties [53–55]. For this reason, after total replacement of the sand volume, a decrease was observed of the mechanical strengths of the conglomerates, this effect is ascribable to the low density/high porosity of the EPS beads (inset Figure 2) and to the voids created by the aggregate at the cement/EPS interface during mixing [53,54]. In fact, the porosity of these samples is approximately two times higher than the references (Table 2). For this purpose, the flexural and the compressive resistances of EPS2, EPS3, and EPS4 samples were approximately ~80% lower than the references, with compressive strengths passing from nearly 50 MPa to less than 10 MPa as the specific mass was lowered from 2100 to 900 kg/m^3. After substitution of 50% of the sand volume with EPS beads (Sand-EPS), an increase was observed of the mechanical strengths with respect to the EPS specimens. In fact, the flexural strength decrease was approximately 25% with respect to both references, while the compressive strength was 25%–30% lower than the references.

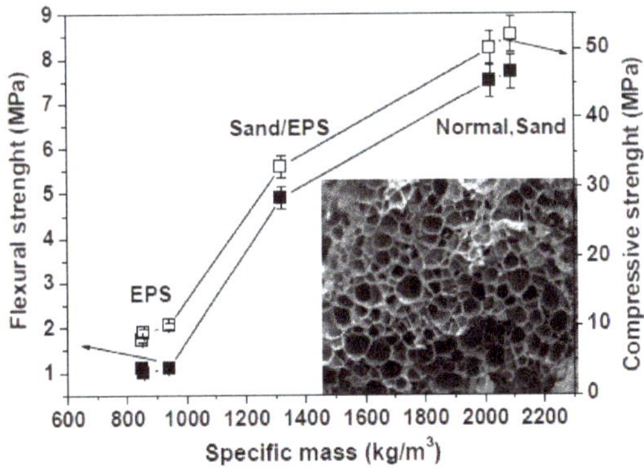

Figure 2. Flexural and compressive strengths of the samples (28 days curing). The label EPS (expanded polystyrene) represents EPS 2, EPS3, and EPS4. White squares represent the compressive strengths, while black squares represent flexural strengths. In the inset: inner porosity of an EPS bead (SEM image).

The EPS mortars did not show a flexural brittle behavior which can be observed in the sand specimens (Normal and Sand), but the rupture was more gradual and the mortars containing 100% EPS volume did not show a separation of the two parts [56,57]. The Sand-EPS sample, characterized by 50% of sand and 50% of EPS, showed a semi-brittle behavior. As in the former case, the compressive failure of the EPS2, EPS3, and EPS4 mortars was gradual with high energy absorption because of the load retention after rupture without collapse [56,58,59]. As expected, the reference samples showed a typical brittle failure. It was observed that most of the aggregates of the EPS3 and EPS4 specimens sheared off along the failure plane (Figure 3A,B), on the contrary no damage was observed to most of the aggregates in EPS2 mortar and some of the EPS2 beads were de-bonded from the matrix (Figure 3C).

Table 3. Mechanical strengths (28 days curing) of the samples.

Sample	ρ (Kg/m³)	R_F (MPa)	R_C (MPa)
Normal	2020	7.5	50
Sand	2090	7.7	52
Sand-EPS	1320	4.9	33
EPS 2	850	1.1	8
EPS 3	940	1.1	10
EPS 4	855	1	9

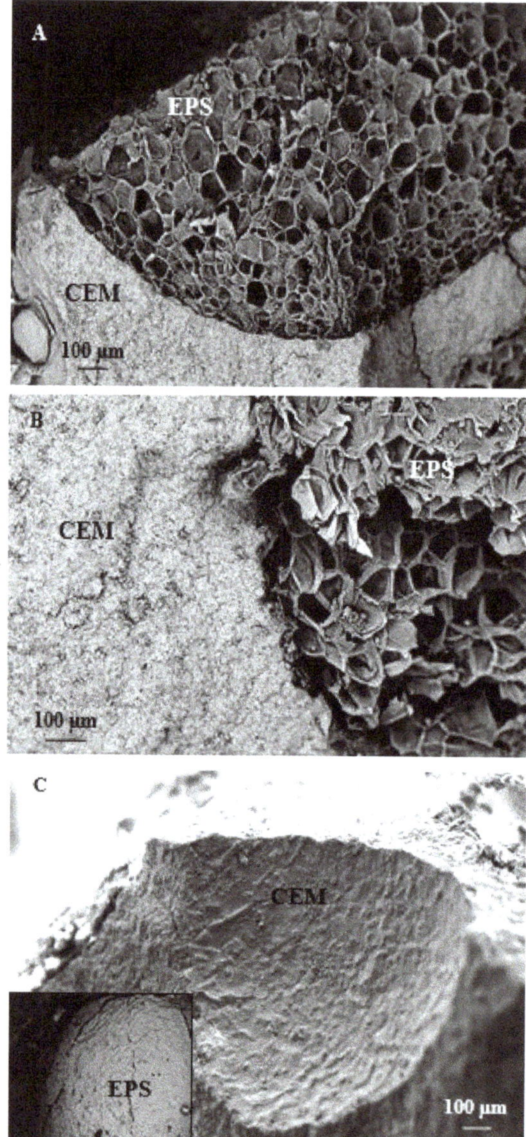

Figure 3. (**A**) SEM image of the cement paste/EPS interface in the EPS3 sample. (**B**) SEM image of the cement paste/EPS interface in the EPS4 sample. (**C**) SEM image of the cement paste/EPS interface in the EPS2 sample, in the inset an image of the de-bonded EPS bead.

From these results, it can be concluded that the bond between the EPS2 aggregate and the cement paste was weaker than the failure strength of the aggregate (poor EPS adhesion to the cement paste), while the bond between the EPS aggregate and the cement paste in EPS3 and EPS4 samples was stronger (better EPS adhesion to the cement paste) than the failure strength of the polystyrene granules [33,60]. This effect was noticed in particular on the EPS3 sample (Figure 3A). The latter result is indicative of better cohesion between the aggregate and the cement paste. Thus, EPS3 exhibited higher compaction which packs the aggregate particles together, so as to increase the specific mass of the mortar, and this also explains the lower percentage flow with respect to the other samples which resulted in more fluid and with a higher tendency to segregation [20] (see Figure 1).

The lower specific mass of the EPS2 sample can be demonstrated by large voids at the ligand/aggregate interface, with length comparable to EPS beads and 20–30 micron width, this effect was ascribed to the mentioned poor adhesion of the beads to the cement paste (Figure 4A,B). This result was also observed in the EPS3 sample, but in the latter case the adhesion of the sheared-off particles to the cement paste was better thus demonstrating the higher specific mass of this type of lightweight mortar. Moreover, from Figure 4C, the perfect adhesion of the sand to the cement paste is evident. In fact, from the map relative to the Si element, which is barely present in the limestone, a negligible separation between the sand and the ligand can be observed ascribed to a favorable adhesion.

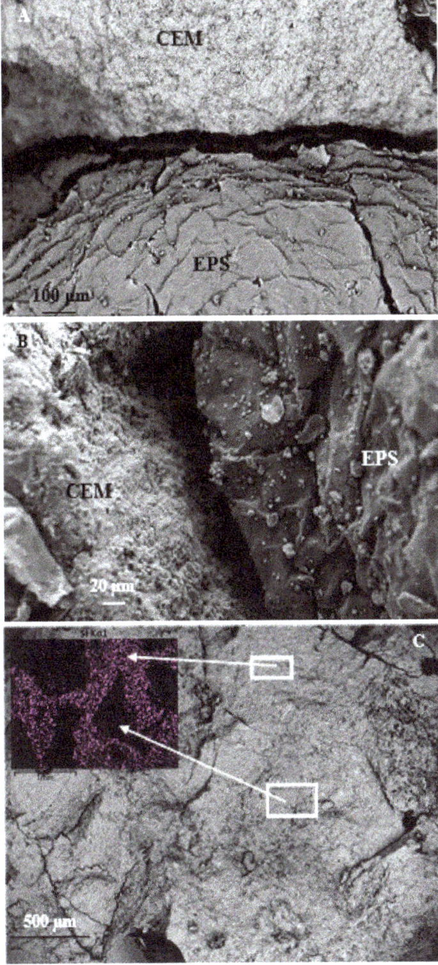

Figure 4. (**A**,**B**) SEM images of the cement paste/EPS interface in the EPS2 sample. (**C**) SEM image of the normalized mortar and, in the inset, EDX map relative to the Si distribution in the sample.

The time variation of the flexural and compressive strengths of the Normal sample, of the EPS3 and Sand/EPS samples is reported in Figure 5 where an increase of the resistances can be observed with stabilization after 45 days. At 60 days, the values did not sensibly change thus demonstrating stability of the materials on consideration of the specific water curing/conservation conditions of the conglomerates.

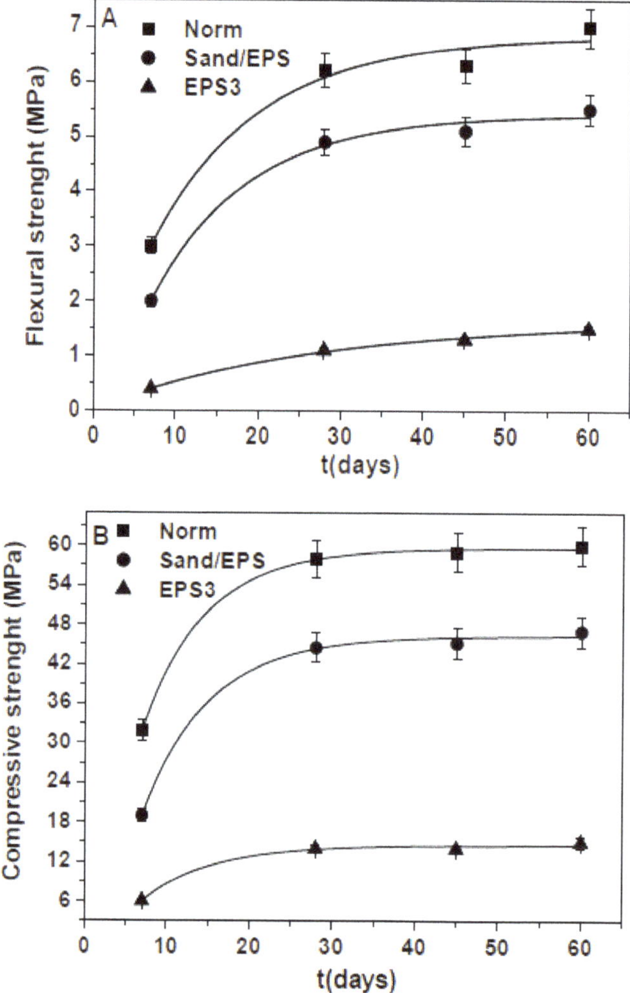

Figure 5. Flexural (**A**) and compressive (**B**) strengths of the samples over time.

EPS based mortars showed lower thermal conductivities and diffusivities than the sand references (Figure 6). This result can be ascribed to the lower specific mass of the specimens due to the low density of the organic aggregates [61,62] (see inset Figure 2) together with the mentioned voids at the EPS/ligand interface which limit heat transport in the composite. Specifically, the thermal conductivities of the bare EPS specimens were ~80% lower than the references. The best results were obtained in the case of the EPS4 specimen (0.29 W/mK) due to the lowest specific mass. Intermediate values (0.8 W/mK) were obtained in samples with 50% of EPS (Sand/EPS sample). Thermal conductivity and diffusivity data showed an exponential decrease with the decrease of the conglomerates specific mass.

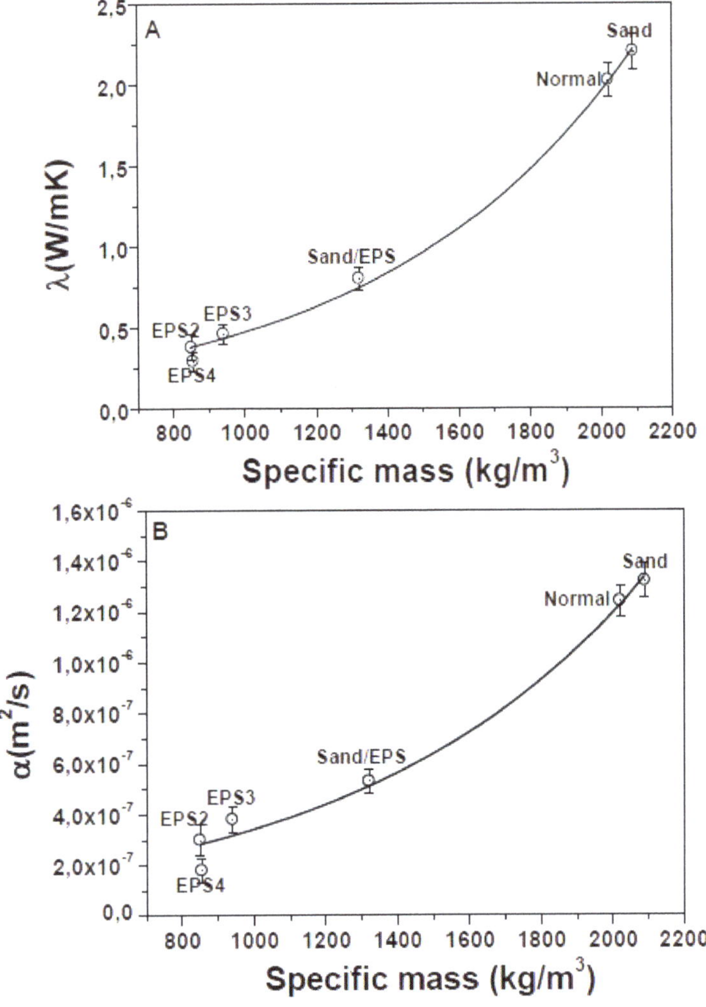

Figure 6. (**A**) Thermal conductivity and (**B**) thermal diffusivity of the samples.

The wetting characterization of the side surface (Figure 7) and of the inner surface (Figure 8) of the Normal sample was carried out. Figure 7A,B shows the time evolution of the water contact angle (WCA) and of the drop height for the side surface of the Sand sample. A hydrophilic character (WCA < 90°) [35] was observed although different behavior on various points of observation was detected. Fast WCA decrease and full penetration occurred in few seconds at point 3, slower but full water absorption occurred at point 2, whereas higher WCA and negligible water absorption were observed in the case of point 1. Figure 7C shows the pictures relative to the drop behavior. The side surface of the reference mortar based on normalized sand (Normal) showed similar features. It is worth highlighting that the possibility of detecting and quantifying the spatially non-homogeneous behavior of a surface/material like these is a specific benefit of the spatially resolved evaluation of wettability and absorption made by this technique (the drop volume is 5 μL), which cannot be achieved with water permeability or capillary absorption measurements.

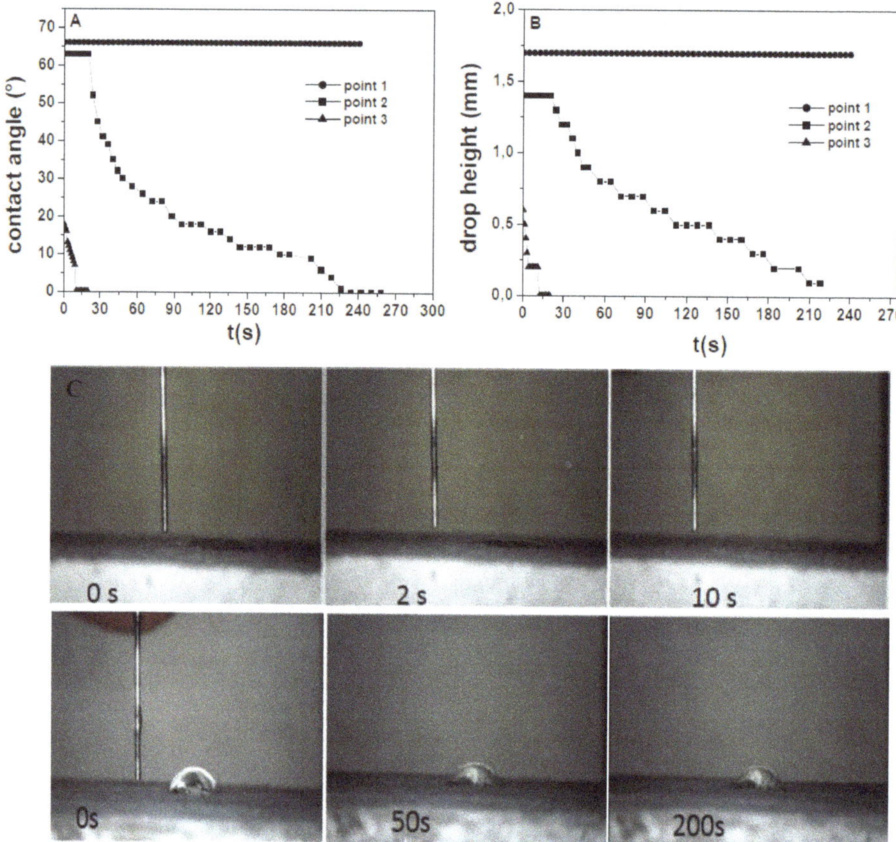

Figure 7. (**A**) Contact angle and (**B**) height variation over time for water drops deposited on representative points of the side surface of the normalized mortar (Sand). (**C**) Optical microscope images (down: point 1 drop, top: point 3 drop).

Figure 8A,B shows the wetting parameters relative to the fracture surface. The inner surface resulting from the mechanical breakage can be considered more representative of the composite features because it is a section of the sample showing every component of the mixture. It shows an open porosity characterized by high roughness and a visible distribution of the aggregates, as opposed to what is observed on the side surface. Specifically, the results obtained on every point of the observation were similar. A fast decrease of the water contact angle and of the drop height was observed at every point (Figure 8C). On the contrary to what was observed on the side surface, WCA was lower, thus the fracture surface can be generally considered super-hydrophilic (WCA ~0–5 [35,63] and fast absorbent. As in the former case, similar results were observed on the inner surface of the Normal sample.

The wetting characterization of the EPS3 mortar, with EPS grains in the 2–4 mm (50%) and 4–6 mm (50%) bead size range, is reported in Figures 9 and 10. As described above, EPS totally replaced the sand volume. Figure 9A,B shows the time evolution of the water contact angle (WCA) and of the drop height on the side surface of the sample. Different trends were observed. A slow but full water absorption occurred at point 1, higher WCA and negligible water absorption were observed in the case of points 2 and 3, the latter with WCA $\geq$ 90°. In the present case, the side surface resulted in being more hydrophobic than the references.

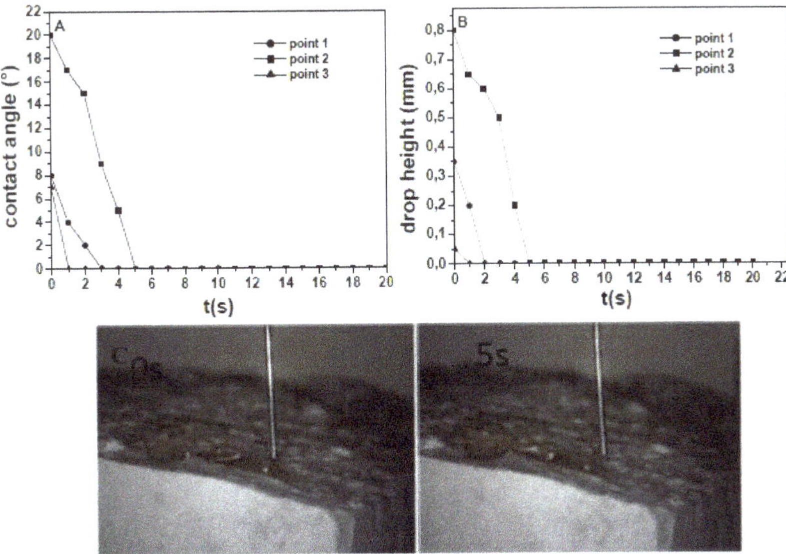

Figure 8. (**A**) Contact angle and (**B**) drop height for representative points of the fracture surface of the normalized mortar (Sand). (**C**) In the optical microscope image: point 2 drop.

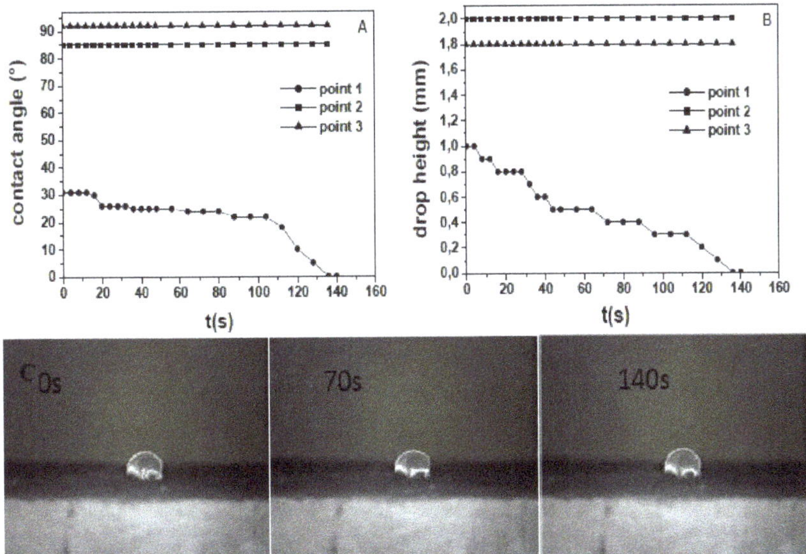

Figure 9. (**A**) Contact angle and (**B**) drop height for representative points of the side surface of the EPS3 mortar. (**C**) In the optical microscope image: point 2 drop.

Figure 10A,B shows the time evolution of the water contact angle (WCA) and of the drop height on the fracture surface of the EPS3 sample. In this case, the drop was stable for all the observation time. Figure 10 also shows a picture of the drop after deposition onto the specimen surface (point 2) which resulted in being hydrophobic with high WCA (WCA > 90°) [35]. The latter result was confirmed after a drop deposition onto an EPS slab or onto bare EPS beads, specifically in the first case WCA was approximately 99°, while higher in the second (100–102°) probably due to the beads' curvature. The WCA was higher on bare beads with respect to the EPS in the mixture because of the absence of contamination from the hydrophilic cement paste [64,65]. For this purpose, after deposition onto cement paste regions of the EPS3 sample (points 1 and 3), hydrophilic behavior but negligible water absorption were observed. This latter result is ascribed to the hydrophobic and non- absorbing effect of EPS whose sites decrease the mean surface energy of the sample making the presence of the porous and hydrophilic cement regions ineffective [64,65].

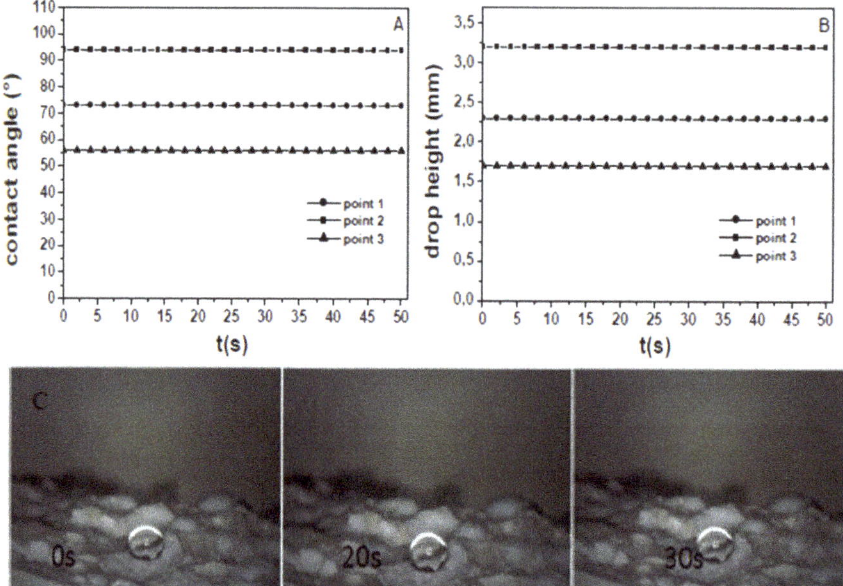

Figure 10. (**A**) Contact angle and (**B**) drop height for representative points of the fracture surface of the EPS3 mortar. (**C**) In the optical microscope image: point 2 drop.

The wetting characterization of the fracture surface of the EPS4 mortar, with EPS grains in the 1–2 mm (25%), 2–4 mm (25%), and 4–6 mm (50%) bead size range, is reported in Figure 11A, while the results obtained on the side surface were similar to those of the EPS3 specimen. The fracture surface is hydrophobic in the domain of the polystyrene beads (point 2) and hydrophilic in the domain of the cement paste (point 3) because the drop was deposited onto a hydrophilic and absorbent surface. As a matter of fact, the latter result represents a difference between the fracture surface of this sample and the fracture surface of the former composite (EPS3).

Wetting characterization of the fracture surface of the EPS2 mortar, with EPS grains in the 4–6 mm (100%) bead size range, is reported in Figure 11B and in this case, the results obtained on the side surface of this sample were similar to those observed in the case of the former EPS specimens. In the case of the fracture surface, a hydrophilic character was observed at every point of observation, with very low water contact angle and fast water absorption.

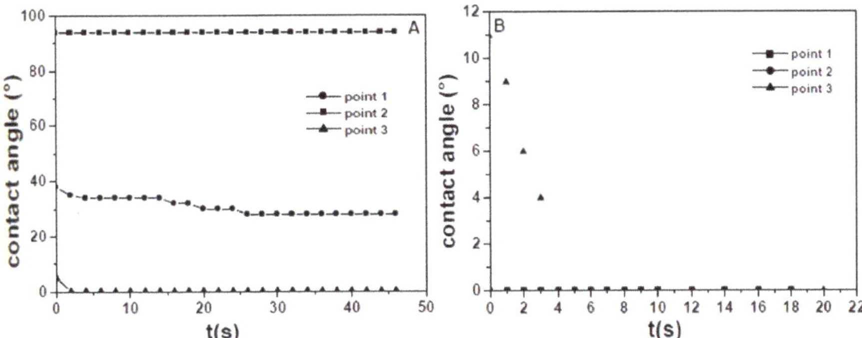

Figure 11. Contact angle for representative points of the fracture surface of (**A**) EPS4 and (**B**) EPS2 mortars.

Thus, EPS3 is the specimen with the lowest water drop absorption. This may be due to a more efficient organization of the aggregate particles with open spaces (spheroidal microcavities) between larger particles filled with smaller size EPS beads [49,66], which leads to better behavior of the composite. This specimen indeed shows the highest specific mass and lowest porosity among the EPS specimens, reasonably as a consequence of a better aggregate compaction (evidenced by the lowest flow). This property on the one hand results in a slight reduction of the thermal insulation performances, but on the other hand makes the composite definitely less subject to water ingress. The importance of optimizing the level of compaction, by adjusting the size distribution of the EPS aggregates, is due to the relatively large size of the initial EPS beads, which results in the formation of too large channels of the cement matrix among the aggregates in the hardened artifacts.

Hence, once properly distributed in size, EPS beads can represent suitable aggregates in cement-based artifacts both for lightening/insulating and water proofing purposes. Such a double advantage arises from the peculiar combination of low density and low surface energy of this plastic matter as already shown by using other polymeric aggregates, such as granulated rubber from end-of-life tires [53].

4. Conclusions

In the present work an investigation into the rheological, thermo-mechanical, microstructural and wetting characteristics of cement mortars containing recycled expanded polystyrene (EPS) was carried out. The samples were prepared after partial/total replacement of the conventional sand aggregate with EPS having different grain size and size distribution. The experimental results may be summarized as follows:

- EPS samples resulted in having more fluid than the references, in particular the sample characterized by EPS grains in the 2–4 mm (50%) and 4–6 mm (50%) bead size range (EPS3) had the most plastic with good particle distribution and cohesion between the ligand and the organic aggregates as also observed in microstructural and porosimetric detections.
- The mechanical resistances of the EPS samples were lower with respect to the controls due to the lower specific mass. An increase of the strengths was observed with a stabilization after 45 days. At 60 days the values did not sensibly change thus demonstrating stability of the materials in consideration of the specific water curing/conservation conditions of the conglomerates.
- The EPS based mortars showed lower thermal conductivities and diffusivities as compared to the references based on sand due to a lower density ascribed to the low density of the aggregates and to the spaces at the EPS/cement paste interface.
- Interesting results in term of high mechanical resistances and low thermal conductivity were obtained in the case of sand-EPS mixtures.

- Reference sand-based mortars showed hydrophilicity (low WCA) and high water penetration, particularly on the fracture surface of the composites, conversely to what was observed in the case of EPS samples which were generally more hydrophobic and less water absorbent. The best results (high WCA and negligible water penetration on the fracture surface) were obtained with the EPS3 sample. This property was ascribed to the low surface energy of the organic aggregate combined with its better particle distribution and compaction within the hydrophilic domains of the cement paste in the composite.
- These lightweight thermo-insulating composites may be used in the building industry as non-structural components, with specific reference to indoor applications (panels, plasters). Moreover, the conglomerates can be considered environmentally sustainable because they are prepared with secondary raw materials (recycled EPS) and are cost-effective because a cheap preparation route was used since the renewable aggregates were not pre-treated and a complex technique of production was not required.

Author Contributions: Conceptualization, A.P.; methodology, A.P.; software, R.D.M.; validation, A.P., R.D.M. and M.N.; formal analysis, A.P.; investigation, A.P., R.D.M.; resources, A.P.; data curation, A.P.; writing—original draft preparation, A.P.; writing—review and editing, A.P., R.D.M., M.N.; visualization, M.N.; supervision, M.N. All authors have read and agreed to the published version of the manuscript.

Funding: This research received no external funding.

Acknowledgments: Special thanks to Pietro Stefanizzi and Stefania Liuzzi for thermal analysis. Adriano Boghetich is acknowlwdged for SEM-EDX analysis and also Regione Puglia (Micro X-Ray Lab Project–Reti di Laboratori Pubblici di Ricerca, cod. n. 45 and 56). Acknowledgments to the DICATECh of the Polytechnic of Bari for SEM analyses.

Conflicts of Interest: The authors declare no conflict of interest.

References

1. Garcia, D.; You, F. Systems engineering opportunities for agricultural and organic waste management in the food–water–energy nexus. *Curr. Opin. Chem. Eng.* **2017**, *18*, 23–31. [CrossRef]
2. Sengupta, A.; Gupta, N.K. MWCNTs based sorbents for nuclear waste management: A review. *J. Environ. Chem. Eng.* **2017**, *5*, 5099–5114. [CrossRef]
3. Li, M.; Liu, J.; Han, W. Recycling and management of waste lead-acid batteries: A mini-review. *Waste Manag. Res.* **2016**, *34*, 298–306. [CrossRef]
4. Asefi, H.; Lim, S. A novel multi-dimensional modeling approach to integrated municipal solid waste management. *J. Clean. Prod.* **2017**, *166*, 1131–1143. [CrossRef]
5. Liuzzi, S.; Rubino, C.; Stefanizzi, P.; Petrella, A.; Boghetich, A.; Casavola, C.; Pappalettera, G. Hygrothermal properties of clayey plasters with olive fibers. *Constr. Build. Mater.* **2018**, *158*, 24–32. [CrossRef]
6. Coppola, L.; Bellezze, T.; Belli, A.; Bignozzi, M.C.; Bolzoni, F.; Brenna, A.; Cabrini, M.; Candamano, S.; Cappai, M.; Caputo, D.; et al. Binders alternative to Portland cement and waste management for sustainable construction-part 1. *J. Appl. Biomater. Funct. Mater.* **2018**, *16*, 186–202.
7. Coppola, L.; Bellezze, T.; Belli, A.; Bignozzi, M.C.; Bolzoni, F.; Brenna, A.; Cabrini, M.; Candamano, S.; Cappai, M.; Caputo, D.; et al. Binders alternative to Portland cement and waste management for sustainable construction-part 2. *J. Appl. Biomater. Funct. Mater.* **2018**, *16*, 207–221.
8. Ossa, A.; García, J.L.; Botero, E. Use of recycled construction and demolition waste (CDW) aggregates: A sustainable alternative for the pavement construction industry. *J. Clean. Prod.* **2016**, *135*, 379–386. [CrossRef]
9. Gómez-Meijide, B.; Pérez, I.; Pasandín, A.R. Recycled construction and demolition waste in cold asphalt mixtures: Evolutionary properties. *J. Clean. Prod.* **2016**, *112*, 588–598. [CrossRef]
10. Petrella, A.; Cosma, P.; Rizzi, V.; De Vietro, N. Porous alumosilicate aggregate as lead ion sorbent in wastewater treatments. *Separations* **2017**, *4*, 25. [CrossRef]
11. Xuan, D.X.; Molenaar, A.A.A.; Houben, L.J.M. Evaluation of cement treatment of reclaimed construction and demolition waste as road bases. *J. Clean. Prod.* **2015**, *100*, 77–83. [CrossRef]

12. Petrella, A.; Petruzzelli, V.; Ranieri, E.; Catalucci, V.; Petruzzelli, D. Sorption of Pb(II), Cd(II) and Ni(II) from single- and multimetal solutions by recycled waste porous glass. *Chem. Eng. Commun.* **2016**, *203*, 940–947. [CrossRef]

13. Petrella, A.; Petrella, M.; Boghetich, G.; Basile, T.; Petruzzelli, V.; Petruzzelli, D. Heavy metals retention on recycled waste glass from solid wastes sorting operations: A comparative study among different metal species. *Ind. Eng. Chem. Res.* **2012**, *51*, 119–125. [CrossRef]

14. Petrella, A.; Petruzzelli, V.; Basile, T.; Petrella, M.; Boghetich, G.; Petruzzelli, D. Recycled porous glass from municipal/industrial solid wastes sorting operations as a lead ion sorbent from wastewaters. *React. Funct. Polym.* **2010**, *70*, 203–209. [CrossRef]

15. Singh, N.; Hui, D.; Singh, R.; Ahuja, I.P.S.; Feo, L.; Fraternali, F. Recycling of plastic solid waste: A state of art review and future applications. *Compos. Part B Eng.* **2017**, *115*, 409–422. [CrossRef]

16. Lopez, G.; Artetxe, M.; Amutio, M.; Alvarez, J.; Bilbao, J.; Olazar, M. Recent advances in the gasification of waste plastics: A critical overview. *Renew Sustain. Energy Rev.* **2018**, *82*, 576–596. [CrossRef]

17. Lopez, G.; Artetxe, M.; Amutio, M.; Bilbao, J.; Olazar, M. Thermochemical routes for the valorization of waste polyolefinic plastics to produce fuels and chemicals: A review. *Renew Sustain. Energy Rev.* **2017**, *73*, 346–368. [CrossRef]

18. Rajaeifar, M.A.; Abdi, R.; Tabatabaei, M. Expanded polystyrene waste application for improving biodiesel environmental performance parameters from life cycle assessment point of view. *Renew Sustain. Energy Rev.* **2017**, *74*, 278–298. [CrossRef]

19. Maharana, T.; Negi, Y.S.; Mohanty, B. Review article: Recycling of polystyrene. *Polym. Plast. Technol. Eng.* **2007**, *46*, 729–736. [CrossRef]

20. Herki, B. Combined effects of densified polystyrene and unprocessed fly ash on concrete engineering properties. *Buildings* **2017**, *7*, 77. [CrossRef]

21. Bayoumi, T.A.; Tawfik, M.E. Immobilization of sulfate waste simulate in polymer–cement composite based on recycled expanded polystyrene foam waste: Evaluation of the final waste form under freeze–thaw treatment. *Polym. Compos.* **2017**, *38*, 637–645. [CrossRef]

22. Song, H.Y.; Cheng, X.X.; Chu, L. Effect of density and ambient temperature on coefficient of thermal conductivity of heat-insulated EPS and PU materials for food packaging. *Appl. Mech. Mater.* **2014**, *469*, 152–155. [CrossRef]

23. Loddo, V.; Marcì, G.; Palmisano, G.; Yurdakal, S.; Brazzoli, M.; Garavaglia, L.; Palmisano, L. Extruded expanded polystyrene sheets coated by TiO_2 as new photocatalytic materials for foodstuffs packaging. *Appl. Surf. Sci.* **2012**, *261*, 783–788. [CrossRef]

24. Cai, S.; Zhang, B.; Cremaschi, L. Review of moisture behavior and thermal performance of polystyrene insulation in building applications. *Build. Environ.* **2017**, *123*, 50–65. [CrossRef]

25. Haibo, L. Experimental study on preparation of fly ash polystyrene new insulation building material. *Chem. Eng. Trans.* **2017**, *59*, 295–300.

26. Khoukhi, M.; Fezzioui, N.; Draoui, B.; Salah, L. The impact of changes in thermal conductivity of polystyrene insulation material under different operating temperatures on the heat transfer through the building envelope. *Appl. Therm. Eng.* **2016**, *105*, 669–674. [CrossRef]

27. Patiño-Herrera, R.; Catarino-Centeno, R.; González-Alatorre, G.; Gama Goicochea, A.; Pérez, E. Enhancement of the hydrophobicity of recycled polystyrene films using a spin coating unit. *J. Appl. Polym. Sci.* **2017**, *134*, 45365. [CrossRef]

28. Mohajerani, A.; Ashdown, M.; Abdihashi, L.; Nazem, M. Expanded polystyrene geofoam in pavement construction. *Constr. Build. Mater.* **2017**, *157*, 438–448. [CrossRef]

29. Tawfik, M.E.; Eskander, S.B.; Nawwar, G.A.M. Hard wood-composites made of rice straw and recycled polystyrene foam wastes. *J. Appl. Polym. Sci.* **2017**, *134*, 44770. [CrossRef]

30. Kim, H.; Park, S.; Lee, S. Acoustic performance of resilient materials using acrylic polymer emulsion resin. *Materials* **2016**, *9*, 592. [CrossRef]

31. Dissanayake, D.M.K.W.; Jayasinghe, C.; Jayasinghe, M.T.R. A comparative embodied energy analysis of a house with recycled expanded polystyrene (EPS) based foam concrete wall panels. *Energy Build.* **2017**, *135*, 85–94. [CrossRef]

32. Herki, B.A.; Khatib, J.M. Valorisation of waste expanded polystyrene in concrete using a novel recycling technique. *Eur. J. Environ. Civ. Eng.* **2017**, *21*, 1384–1402. [CrossRef]

33. Babu, D.S.; Ganesh Babu, K.; Tiong-Huan, W. Effect of polystyrene aggregate size on strength and moisture migration characteristics of lightweight concrete. *Cem. Concr. Compos.* **2006**, *28*, 520–527. [CrossRef]

34. Fernando, P.L.N.; Jayasinghe, M.T.R.; Jayasinghe, C. Structural feasibility of Expanded Polystyrene (EPS) based lightweight concrete sandwich wall panels. *Constr. Build. Mater.* **2017**, *139*, 45–51. [CrossRef]

35. Sethi, S.K.; Manik, G. Recent progress in super hydrophobic/hydrophilic self-cleaning surfaces for various industrial applications: A review. *Polym. Plast. Technol.* **2018**, *57*, 1932–1952. [CrossRef]

36. Di Mundo, R.; Bottiglione, F.; Carbone, G. Cassie state robustness of plasma generated randomly nano-rough surfaces. *Appl. Surf. Sci.* **2014**, *16*, 324–332. [CrossRef]

37. Di Mundo, R.; D'Agostino, R.; Palumbo, F. Long-lasting antifog plasma modification of transparent plastics. *ACS Appl. Mater. Interfaces* **2014**, *6*, 17059–17066. [CrossRef]

38. Di Mundo, R.; Dilonardo, E.; Nacucchi, M.; Carbone, G.; Notarnicola, M. Water absorption in rubber-cement composites: 3D structure investigation by X-ray computed tomography. *Constr. Build. Mater.* **2019**, *228*, 116602. [CrossRef]

39. Yue, P.; Renardy, Y. Spontaneous penetration of a non-wetting drop into an exposed pore. *Phys. Fluids* **2013**, *25*, 052104. [CrossRef]

40. Neto, E.; Magina, S.; Camoes, A.; Cachim, L.P.; Begonha, A.; Evtuguin, D.V. Characterization of concrete surface in relation to graffiti protection coatings. *Constr. Build. Mater.* **2016**, *102*, 435–444. [CrossRef]

41. Weisheit, S.; Unterberger, S.H.; Bader, T.; Lackner, R. Assessment of test methods for characterizing the hydrophobic nature of surface-treated High Performance Concrete. *Constr. Build. Mater.* **2016**, *110*, 145–153. [CrossRef]

42. European Committee for Standardization. Products and Systems for the Protection and Repair of Concrete Structures. Definitions, Requirements, Quality Control and Evaluation of Conformity in Part 2: Surface Materials and Structures Protection Systems for Concretes. BS EN 1504-2. Available online: \unhbox\voidb@ x\hbox{https://shop.bsigroup.com/ProductDetail/?pid=000000000030036789} (accessed on 21 July 2019).

43. Ramachandran, R.; Sobolev, K.; Nosonovsky, M. Dynamics of droplet impact on hydrophobic/icephobic concrete with the potential for superhydrophobicity. *Langmuir* **2015**, *31*, 1437–1444. [CrossRef] [PubMed]

44. Flores-Vivian, I.; Hejazi, V.; Khozhukova, M.I.; Nosonovsky, M.; Sobolev, K. Self-assembling particle-siloxane coatings for superhydrophobic concrete. *ACS Appl. Mater. Interfaces* **2013**, *5*, 13284–13294. [CrossRef] [PubMed]

45. Italian Organization for Standardization (UNI). Cement Composition, Specifications and Conformity Criteria for Common Cements. EN 197-1. Available online: http://store.uni.com/magento-1.4.0.1/index.php/en-197-1-2011.html (accessed on 14 September 2011).

46. Italian Organization for Standardization (UNI). Methods of Testing Cement-Part 1: Determination of Strength. EN 196-1. Available online: http://store.uni.com/magento-1.4.0.1/index.php/en-196-1-2016.html (accessed on 27 April 2016).

47. International Organization for Standardization (ISO). Cement, Test Methods, Determination Of Strength. ISO 679. Available online: http://store.uni.com/magento-1.4.0.1/index.php/iso-679-2009.html (accessed on 24 April 2009).

48. Petrella, A.; Spasiano, D.; Rizzi, V.; Cosma, P.; Race, M.; De Vietro, N. Lead ion sorption by perlite and reuse of the exhausted material in the construction field. *Appl. Sci.* **2018**, *8*, 1882. [CrossRef]

49. Petrella, A.; Petrella, M.; Boghetich, G.; Petruzzelli, D.; Ayr, U.; Stefanizzi, P.; Calabrese, D.; Pace, L. Thermo-acoustic properties of cement-waste-glass mortars. *Proc. Inst. Civ. Eng. Constr. Mater.* **2009**, *162*, 67–72. [CrossRef]

50. Petrella, A.; Spasiano, D.; Acquafredda, P.; De Vietro, N.; Ranieri, E.; Cosma, P.; Rizzi, V.; Petruzzelli, V.; Petruzzelli, D. Heavy metals retention (Pb (II), Cd (II), Ni (II)) from single and multimetal solutions by natural biosorbents from the olive oil milling operations. *Process Saf. Environ. Prot.* **2018**, *114*, 79–90. [CrossRef]

51. Italian Organization for Standardization (UNI). Determination of Consistency of Cement Mortars Using a Flow Table. 7044. Available online: http://store.uni.com/magento-1.4.0.1/index.php/uni-7044-1972.html (accessed on 20 April 1972).

52. Gustafsson, S.E. Transient plane source techniques for thermal conductivity and thermal diffusivity measurements of solid materials. *Rev. Sci. Instrum.* **1991**, *62*, 797–804. [CrossRef]

53. Di Mundo, R.; Petrella, A.; Notarnicola, M. Surface and bulk hydrophobic cement composites by tyre rubber addition. *Constr. Build. Mater.* **2018**, *172*, 176–184. [CrossRef]

54. Petrella, A.; Spasiano, D.; Liuzzi, S.; Ayr, U.; Cosma, P.; Rizzi, V.; Petrella, M.; Di Mundo, R. Use of cellulose fibers from wheat straw for sustainable cement mortars. *J. Sustain. Cem. Based Mater.* **2019**, *8*, 161–179. [CrossRef]

55. Spasiano, D.; Luongo, V.; Petrella, A.; Alfè, M.; Pirozzi, F.; Fratino, U.; Piccinni, A.F. Preliminary study on the adoption of dark fermentation as pretreatment for a sustainable hydrothermal denaturation of cement-asbestos composites. *J. Clean. Prod.* **2017**, *166*, 172–180. [CrossRef]

56. Al-Manaseer, A.A.; Dalal, T.R. Concrete containing plastic aggregates. *Concr. Int.* **1997**, *19*, 47–52.

57. Li, G.; Stubblefield, M.A.; Garrick, G.; Eggers, J.; Abadie, C.; Huang, B. Development of waste tire modified concrete. *Cem. Concr. Res.* **2004**, *34*, 2283–2289. [CrossRef]

58. Ganesh Babu, K.; Saradhi Babu, D. Behaviour of lightweight expanded polystyrene concrete containing silica fume. *Cem. Concr. Res.* **2003**, *33*, 755–762. [CrossRef]

59. Saradhi Babu, D.; Ganesh Babu, K.; Wee, T.H. Properties of lightweight expanded polystyrene aggregate concretes containing fly ash. *Cem. Concr. Res.* **2005**, *35*, 1218–1223. [CrossRef]

60. Laukaitis, A.; Zurauskas, R.; Keriene, J. The effect of foam polystyrene granules on cement composite properties. *Cem. Concr. Compos.* **2005**, *27*, 41–47. [CrossRef]

61. Petrella, A.; Spasiano, D.; Rizzi, V.; Cosma, P.; Race, M.; De Vietro, N. Thermodynamic and kinetic investigation of heavy metals sorption in packed bed columns by recycled lignocellulosic materials from olive oil production. *Chem. Eng. Commun.* **2019**, *206*, 1715–1730. [CrossRef]

62. Petrella, A.; Spasiano, D.; Race, M.; Rizzi, V.; Cosma, P.; Liuzzi, S.; De Vietro, N. Porous waste glass for lead removal in packed bed columns and reuse in cement conglomerates. *Materials* **2019**, *12*, 94. [CrossRef]

63. Giannuzzi, G.; Gaudioso, C.; Di Mundo, R.; Mirenghi, L.; Fraggelakis, F.; Kling, R.; Lugarà, P.M.; Ancona, A. Short and long term surface chemistry and wetting behaviour of stainless steel with 1D and 2D periodic structures induced by burst of femtosecond laser pulses. *Appl. Surf. Sci.* **2019**, *494*, 1055–1065. [CrossRef]

64. Song, Z.; Xue, X.; Li, Y.; Yang, J.; He, Z.; Shen, S.; Jiang, L.; Zhang, W.; Xu, L.; Zhang, H.; et al. Experimental exploration of the waterproofing mechanism of inorganic sodium silicate-based concrete sealers. *Constr. Build. Mater.* **2016**, *104*, 276–283. [CrossRef]

65. Li, F.; Yang, Y.; Tao, M.; Li, X. A cement paste–tail sealant interface modified with a silane coupling agent for enhancing waterproofing performance in a concrete lining system. *RSC Adv.* **2019**, *9*, 7165–7175. [CrossRef]

66. Petrella, A.; Petrella, M.; Boghetich, G.; Petruzzelli, D.; Calabrese, D.; Stefanizzi, P.; De Napoli, D.; Guastamacchia, M. Recycled waste glass as aggregate for lightweight concrete. *Proc. Inst. Civ. Eng. Constr. Mater.* **2007**, *160*, 165–170. [CrossRef]

Article

Chitosan Biopolymer from Crab Shell as Recyclable Film to Remove/Recover in Batch Ketoprofen from Water: Understanding the Factors Affecting the Adsorption Process

Vito Rizzi [1], Jennifer Gubitosa [2], Paola Fini [2], Roberto Romita [1], Sergio Nuzzo [2] and Pinalysa Cosma [1,2,*]

[1]   Dipartimento di Chimica, Università degli Studi "Aldo Moro", Bari, 4-70126 Bari, Italy; vito.rizzi@uniba.it (V.R.); roberto.romita@uniba.it (R.R.)
[2]   Consiglio Nazionale delle Ricerche CNR-IPCF, UOS Bari, 4-70126 Bari, Italy; j.gubitosa@ba.ipcf.cnr.it (J.G.); p.fini@ba.ipcf.cnr.it (P.F.); sergio.nuzzo@ba.ipcf.cnr.it (S.N.)
*   Correspondence: pinalysa.cosma@uniba.it; Tel.: +39-08-0544-3443

Received: 18 October 2019; Accepted: 18 November 2019; Published: 20 November 2019

Abstract: Seafood, a delight for many people, is sold in the market as a wide variety of products. However, seafood industries produce many by-products; for example, during the processing, the heads and shells of shellfish are generated as waste. This results in the generation of a large amount of shell waste that is accumulated over time, inducing a major environmental concern. Effective solutions for recycling shell waste should be taken into consideration, and the extraction of commercially useful substances like chitin and its derivates, such as chitosan, could be a valid solution for reducing the seafood waste's environmental impact. Thus, during this work, we propose the use of chitosan as biowaste, to induce the formation of solid films useful for decontaminating water from emerging pollutants. In particular, ketoprofen was used as a model contaminant, and a high percentage of removal, at least 90%, was obtained in a short time under our experimental conditions. Thus, a comprehensive investigation into the adsorption of ketoprofen onto chitosan film was performed, detailing the nature of the adsorption by studying the effects of pH, temperature changes, and electrolyte presence in the solutions containing the pollutant. The process was found to be pH-dependent, involving meanly electrostatic interactions between the pollutant molecules and chitosan. The endothermic character of the adsorption was inferred. The kinetics of the process was investigated, showing that the pseudo second-order kinetic model best fit the experimental data. A recycling process of the adsorbent was proposed; therefore, the adsorbed pollutant can be recovered by reusing the same adsorbent material for further consecutive cycles of adsorption without affecting the efficiency for ketoprofen removal from water.

Keywords: chitosan film; emerging pollutants; ketoprofen; food waste; adsorption; recycle

---

1. Introduction

The production of shell waste by the seafood industry is, among other environmental problems, one of noteworthy concern, contributing to environmental and health hazards [1]. Indeed, a great amount of crab and shrimp shells are produced as waste by worldwide seafood companies [2]. Unfortunately, about 45% of waste resulting from processed seafood is disposed as landfill, consequently leading to environmental pollution in terms of odor and aesthetic damage to the environment [3]. To dispose of this waste, burning is proposed; however, this solution is costly due to the low burning capacity of shells [1].

However, it is worth mentioning that fishery by-products have economical value for chitin and chitosan production [3]; thus, the conversion of shell waste to commercial products could be considered an effective approach for shell waste remediation [1]. Accordingly, Premasudha et al. [4] reported that the ocean ecosystem is one of the major sources of biopolymers, particularly chitosan, increasing the interest toward this bio-waste material, which is characterized by very interesting applications due to its chemical structure and properties [4]. Indeed, chitosan, a by-product from the alkaline deacetylation process of chitin [5], is an amino polysaccharide known for its distinctive properties, as well as its biodegradability and biocompatibility [4].

Chitosan has great potential for environmental applications [6], such as the remediation of organic and inorganic contaminants, including toxic metals and dyes in soil, sediment, and water [7–10], and for the development of devices [11,12]. Among pollutants [13,14], emerging contaminants, e.g., pesticides and their metabolites, pharmaceuticals, personal and house care products, life-style compounds, food additives, industrial products and waste, and nanomaterials, are a great and important problem for the environment. The cumulative use of these substances led to their relatively recent appearance in detectable levels in soils, as well as surface and groundwater resources, with unpredictable consequences for ecosystems [15]. The removal of these pollutants from the environment must be taken into account, and the development of a more sustainable and greener technology for this purpose should be developed [16]. Addressing this concern, several studies [17,18] were reported in the literature, with the use of adsorption methods suggested as the most powerful tools. [18] Furthermore, in order to reduce the environmental impact and the associated cost, agricultural and food waste was carefully investigated as a tool for the removal of pollutants [19,20]. Starting from these considerations, with the aim of valorizing food waste, particularly chitosan from crab shell, attention in this work was focused on the removal of ketoprofen (Kp), adopted as a model emerging pollutant. The use of chitosan as a solid film (CH) is proposed, exhibiting enormous advantages in the treatment of water for Kp removal. Kp, a non-steroidal anti-inflammatory drug (NSAID), is frequently found in surface water, constituting a potential risk for aquatic ecosystems [21–23].

Indeed, for its removal, Jankunaite et al. [24] proposed the use of advanced oxidation processes (AOPs). In this regard, however, it is worth mentioning that AOPs suffer from side effects due to the potential toxicity of the induced by-products, which are often more toxic than the parent molecules [24].

Nagy et al. [25] reported the use of cyclodextrin-based polymers, working in synergy with filters, to exploit the possibility of working under dynamic conditions [25]. Madikizela et al. [26] showed the efficiency of molecularly imprinted polymers, designed for the selective extraction of ketoprofen from wastewater [26].

In addition to these studies, other proposed uses of CH to treat water for removal of the emerging contaminant Kp were outlined in our previous paper devoted to the removal of diclofenac [27]. In the present work, a deep investigation into the process is presented. The use of chitosan film from food waste should not only lower the associated adsorption processes, if compared with the previous adsorbents, but also improve the environmental impact of seafood waste, valorizing it with a simple and easy method to obtain materials for treating water with Emerging Contaminants (EC). Moreover, during this work, for the first time, in addition to the possibility of decontaminating water containing Kp, we propose the recovery of CH after adsorption, emphasizing the sustainable character of the described approach. Indeed, 10 consecutive cycles of adsorption/desorption were performed without affecting Kp removal and recovery, which were completed in a few minutes.

2. Materials and Methods

2.1. Chemicals

The used chemicals were of analytical grade, and the samples were prepared using deionized water. Highly viscous chitosan powder, from crab shells (with a molecular weight MW of 150,000 and a deacetylation degree ≥75%, Sigma-Aldrich, Milan, Italy) acetic acid (99.9% Sigma-Aldrich, Milan,

Italy), and glycerol (+99.5%, Sigma-Aldrich, Milan, Italy) were obtained from Sigma-Aldrich. The same commercial source was adopted for sodium ketoprofen ($C_{16}H_{13}NaO_3$, MW: 276.267 $g \cdot mol^{-1}$ Sigma-Aldrich, Milan, Italy), used without further purification. A Kp stock solution with a concentration of 5 $mg \cdot L^{-1}$ was prepared. The pH of the various aqueous solutions, when necessary, was adjusted using concentrated HCl and NaOH solutions. Dilutions were performed using deionized water.

2.2. Preparation of Chitosan Films

Chitosan powder was dissolved in a 0.8% (*v/v*) aqueous acetic acid solution to obtain a 1% (*w/v*) chitosan concentration by continuous stirring for 24 h. Then, 200 µL of glycerol was added to 100 mL of the chitosan/acetic acid solution. The solution was degassed for 1 h and poured into a plastic Petri plate, which was maintained in an oven at 60 °C for 24 h, thereby obtaining CH. The thickness of the obtained film was about 1 mm.

2.3. UV–Visible Light (UV–Vis) Measurements

UV-Vis spectra were recorded using a Varian CARY 5 UV–Vis/near-infrared (NIR) spectrophotometer (Varian Inc., now Agilent Technologies Inc., Santa Clara, CA, USA). Spectra were recorded in the 200–800 nm range, with a 1 nm/s scan rate, and the Kp amounts were monitored by measuring the absorbance intensity at λ = 260 nm. The molar absorption coefficient (ε) of 13,530 $L \cdot mol^{-1} \cdot cm^{-1}$, experimentally inferred, was used to obtain the Kp concentration in water. All experiments were performed in triplicate, calculating the relative standard deviations.

2.4. In Batch Equilibrium Experiments

The CH adsorption capacities were calculated as q_t ($mg \cdot g^{-1}$) at different contact times (t), by applying Equation (1) [13,14,16].

$$q_t = \frac{C_0 - C_t}{W} \times V, \tag{1}$$

where V represents the Kp solution volume (herein 15 mL), W is the dried chitosan adsorbent mass (g), and C_0 and C_t ($mg \cdot L^{-1}$) represent the initial Kp concentration and the Kp concentration at time t.

A fixed amount of CH (150 mg) was added to flasks containing 15 mL of Kp solution at two initial concentrations (1×10^{-5} M and 5×10^{-6} M, corresponding to ~2.5 $mg \cdot L^{-1}$ and ~1.25 $mg \cdot L^{-1}$, respectively). The adsorption was performed under continuous stirring at 250 rpm, and UV–Vis absorption spectra were recorded to evaluate the Kp removal efficiency and the chitosan adsorption capacities from water. Furthermore, the effect of adsorbent amount was also explored, changing its mass from 35 mg to 200 mg. In this case, the Kp concentration was maintained at 1×10^{-5} M. The effects of solution ionic strength and changes in pH values (ranging from 3 to 12) on the adsorption process were also studied.

2.5. In Batch Desorption Experiments

After the Kp adsorption from water, NaCl (0.25 M) was selected, with regard to a green economy and cleaner production, as the best salt to induce the release of the adsorbed Kp. With the same approach adopted for Kp adsorption, the UV–Vis investigation was used to assess the amount of desorbed pollutant. After the Kp adsorption, the adsorbent was washed with fresh water to remove the non-adsorbed Kp swollen in the NaCl solution for release. The effect of the contact time was evaluated, and 30 min was found to be suitable for Kp recovery.

2.6. Adsorption Kinetics

Information about the kinetics of the adsorption process between Kp and the chitosan film was inferred by adopting both pseudo first-order and pseudo second-order kinetic models. The following linearized equations for the pseudo first-order (Equation (2)) and pseudo second-order (Equation (3)) models were adopted [13,14,16]:

$$\ln\left(q_e - q_t\right) = \ln\left(q_e\right) - K_1 \times t,$$ (2)

$$\frac{t}{q_t} = \frac{1}{K_2 q_e^2} + \frac{1}{q_e} \times t,$$ (3)

where q_e and q_t represent the adsorption capacities at equilibrium and at time t, respectively ($mg \cdot g^{-1}$), and k_1 (min^{-1}) and k_2 ($g \cdot mg^{-1} \cdot min^{-1}$) are the rate constants of the pseudo first-order and second-order models, respectively.

2.7. Thermodynamic Studies

Free energy ($\Delta G°$), entropy ($\Delta S°$), and enthalpy ($\Delta H°$) were calculated [27] for the Kp adsorption onto the chitosan film at three selected temperatures of 298, 288, and 278 K. The value of $\Delta G°$ was inferred using Equation (4).

$$\Delta G° = -RT \, ln \, K_{eq},$$ (4)

where R is the universal gas constant (8.314 J/mol·K), T is the temperature (K), and K_{eq} represents the equilibrium constant that was calculated as q_e/C_e, where q_e is the sorption capacity ($mg \cdot g^{-1}$) at equilibrium, and C_e is the equilibrium concentration ($mg \cdot g^{-1}$). The values of $\Delta H°$ and $\Delta S°$ were determined by combining Equation (4) with Equation (5), thereby obtaining Equation (6).

$$\Delta G° = \Delta H° - T\Delta S°.$$ (5)

$$\ln K_{eq} = -\frac{\Delta H°}{RT} + \frac{\Delta S°}{R}.$$ (6)

2.8. Determination of Chitosan Film Zero-Point Charge

The zero-point charge pH (pH_{ZPC}) of the chitosan adsorbent was evaluated by using the pH drift method [14]. Firstly, 30 mL of NaCl solution with a concentration of 5.0×10^{-2} M was used at different pH values ranging from 2 to 12 (pH_i). Concentrated HCl and NaOH solutions were used for this purpose. The pH_i values of these solutions were measured, and 25 mg of adsorbent was subsequently introduced. These solutions were stirred at 298 K for 48 h. The final pH (pH_F) values were measured. By reporting the pH_i versus pH_F values, the value of pH_{ZPC} was inferred at the cross-section of the latter curve with the line of pH_i versus pH_i. All experiments were performed in triplicate, calculating the relative standard deviations.

3. Results and Discussion

The UV–Vis absorption spectrum of Kp was used to monitor its removal from contaminated water. As a first step, an aqueous solution purposely spiked with the drug was investigated, and the Kp spectroscopic main signal at 260 nm (Figure 1A) was followed.

By adopting 150 mg of adsorbent, in the presence of 1×10^{-5} M Kp, the adsorption was followed until 60 min. The main process of Kp removal was observed in the first 15 min, exhibiting an efficiency of 55%; by extending the contact time, the effect resulted less pronounced with respect to the beginning of the process, and, in 60 min, almost 85% of the NSAID was eliminated from the water (Figure 1A). The results suggested the probable key role of the Kp concentration gradient (ΔC) between the bulk of the solution and the adsorbent surface [28]. At the beginning of the adsorption process, the ΔC was high enough to induce the diffusion of the NSAID from the bulk of the solution at the surface of the adsorbent; on the other hand, upon extending the contact time, the ΔC was reduced, thereby slowing down the Kp adsorption. Furthermore, at the beginning of the process, the presence of a large number of free sites on the adsorbent surface for Kp adsorption also favored the NSAID removal. However, upon extending the contact time, the number of available free sites to host Kp decreased, reducing the relative Kp adsorption (Figure 1A) [13,14,16,29].

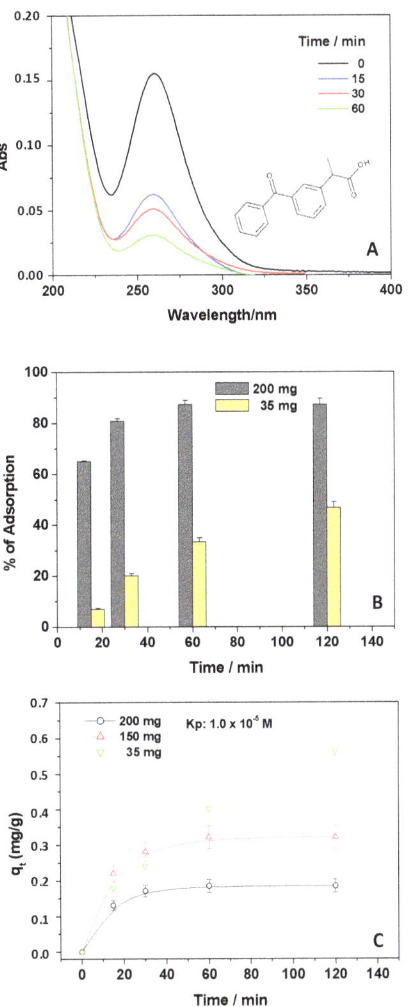

Figure 1. Ultraviolet–visible light (UV–Vis) spectra of a 1×10^{-5} M ketoprofen (Kp) solution, pH 5, collected at different contact times, in the presence of 150 mg of adsorbent (**A**); Kp adsorption percentage (1×10^{-5} M, pH 5) calculated at different contact times in the presence of 200 mg and 35 mg of adsorbent (**B**); adsorption capacities, qt, referring to different amounts of chitosan (200, 150, and 35 mg) in contact with a 1×10^{-5} M Kp solution, pH 5 (**C**).

Moreover, this behavior could be attributed to the presence of repulsive forces between free Kp molecules in solution and those adsorbed, which further hindered the drug adsorption [30].

After these assessments, the effects of several parameters affecting the NSAID removal from water were investigated, while also evaluating the effects on the adsorption capacities.

3.1. Effects of Adsorbent Dosage and Kp Concentration

To study the effect of the CH amount on the Kp removal from water, thus evidencing the role of active sites hosting Kp, two other CH films with different weights (200 and 35 mg) were compared under the same experimental conditions (Figure 1B), i.e., a 1×10^{-5} M Kp solution at pH 5 (pH of the Kp solution soon after adsorbent addition).

The Kp removal efficiencies were calculated, and adsorption percentages of 65% and 8% were determined in the first 15 min when using the greatest and the smallest amounts of adsorbent, respectively; as shown in Figure 1B, these efficiencies were increased upon extending the contact time to 120 min, obtaining 90% Kp removal if in presence of the largest amount of CH.

This result confirmed the previously mentioned importance of free active sites able to host the NSAID [13,14,16,29], which increase in number upon increasing the available surface of the adsorbent. Overall, in Figure 1C, the influence of the chitosan amount on Kp adsorption was evaluated by reporting the q_t values (Equation (1)) as a function of the three investigated CH weights.

The obtained results indicated that, by increasing the adsorbent amount, the relative adsorption of Kp molecules increased (see the plateaus on the graph), while the adsorption capacity decreased. This suggests that, despite the great efficacy, by using a large amount of adsorbent, the adsorption sites remained partially unsaturated during the sorption process, reducing the q_t values as a whole [13,14]. In particular, as already observed by monitoring the Kp UV–Vis absorption spectrum (Figure 1A), at the beginning of the adsorption process (Figure 1C), the presence of a large quantity of free sites for Kp adsorption and the high ΔC increased the q_t values. On the contrary, by extending the contact time, the free sites and ΔC decreased, reducing the Kp adsorption overall, leading to a plateau. All of these findings agree with the results obtained upon varying the Kp concentrations as described below.

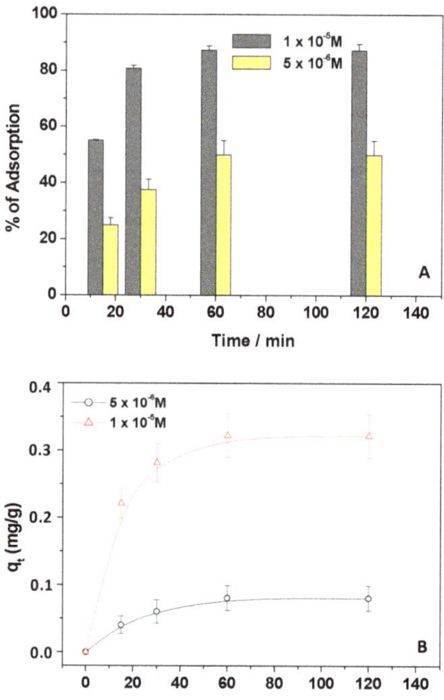

Figure 2. Percentage of Kp adsorption (**A**) and adsorption capacities, qt (**B**), referring to different Kp concentrations of 1×10^{-5} M and 5×10^{-6} M, pH 5, in the presence of 150 mg of adsorbent.

For this purpose, the concentration of Kp was changed (1×10^{-5} M and 5×10^{-6} M), fixing the chitosan weight at 150 mg, as shown in Figure 2A. In both cases, the Kp removal showed great variation in the first minute of contact, showing that, upon diluting the Kp solution, the pollutant removal percentage decreased. Not surprisingly, the dilution of Kp solutions (with a dilution factor of 1:2) reduced the percentage of Kp removal from 65% to 25% in the first 15 min and from 85% to 50% after 120 min. Accordingly, the associated q_t values were calculated, as reported in Figure 2B, showing higher q_t values for the concentrated Kp solution. Moreover, in the latter case, the relative maximum adsorption capacities were quickly obtained after a few min, as compared with the diluted Kp solution.

3.2. Kinetic Analysis

Information about the dynamics of the adsorption was obtained by investigating the kinetics of the process, applying pseudo first-order and pseudo second-order kinetic models (Equations (2) and (3)). By using the q_t values reported in Figure 1B, the results shown in Figure 3A,B were obtained. Table 1 reports the calculated kinetic parameters.

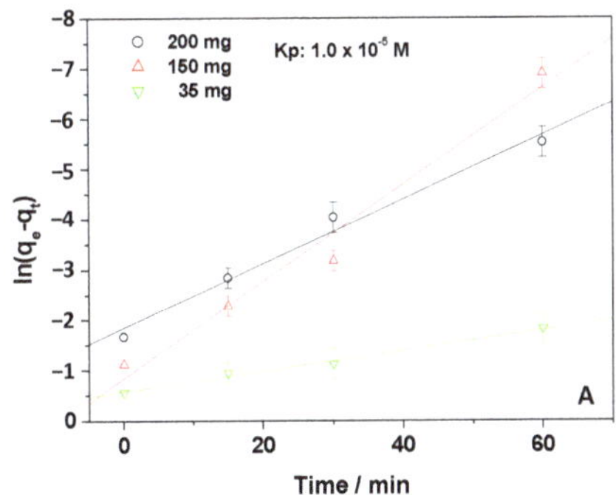

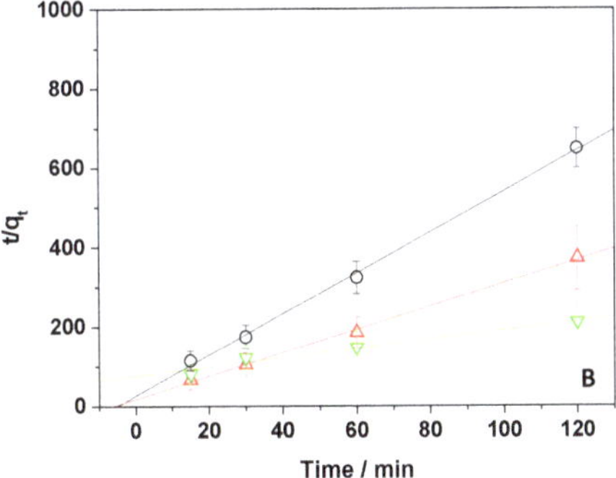

Figure 3. Pseudo first-order (**A**) and second-order (**B**) kinetic models applied to experiments of adsorption in which the amount of the adsorbent was changed.

Table 1. Kinetic parameters obtained by applying the pseudo first-order and second-order kinetic models.

Adsorbent (mg)	Pseudo First-Order				Pseudo Second-Order			
	$q_{e(the)}$	$q_{e(exp)}$	K_1	R^2	$q_{e(the)}$	$q_{e(exp)}$	K_2	R^2
200	0.16	0.18	6.4×10^{-2}	0.983	0.19	0.18	1.05	0.998
150	0.43	0.32	9.6×10^{-2}	0.975	0.33	0.32	0.5	0.998
35	0.55	0.56	2.0×10^{-2}	0.989	0.68	0.56	1.0×10^{-2}	0.982

The best kinetic model to describe the experimental data was evaluated by comparing the R^2 values of the linear fitting, as well as the experimental adsorption capacities at equilibrium, $q_{e,exp}$ (contact time 120 min), with those obtained by applying the kinetic equations, $q_{e,calc}$ (calculated adsorption capacities) [13,14]. From Table 1, the R^2 values and the comparison between $q_{e,exp}$ and $q_{e,calc}$ suggest that the application of the pseudo second-order equation better described the experimental data, emphasizing the role of both Kp and chitosan amounts during the adsorption process [13,14,16]. Antunes et al. [31] reported that the use of this kinetic model indicates that the rate-controlling step depends on both physical and chemical interactions between the pollutant and adsorbent [31]. However, it is worth pointing out that, from the data reported in Table 1, when using the smallest CH amount, the pseudo first-order model is probably preferred. This suggests that, under this condition, the rate limiting step could be the Kp concentration, i.e., the diffusion of Kp mainly controls the removal of the NSAID [28].

Additional information was obtained by adopting the Weber–Morris model (W–M). The intra-particle diffusion model was used, as described by the following equation: $q_t = k_{int} \times t^{1/2} + C$, where C represents the thickness of the boundary layer, and k_{int} is the kinetic constant related to the intra-particle diffusion rate in mg·g^{-1}·min$^{-1/2}$ [32]. As reported by Lin et al. [32], if the plot of q_t versus $t^{1/2}$ is represented by a straight line passing through the origin, the intra-particle diffusion is the limiting stage of the adsorption. On the other hand, if multiple linear segments are necessary to fit the experimental data, two or more steps could be involved during the NSAID adsorption process [31,32]. The W–M equation was, thus, applied (Figure 4A,B) to the q_t values reported in Figures 1C and 2B, referring to experiments in which the amounts of both CH and Kp were changed.

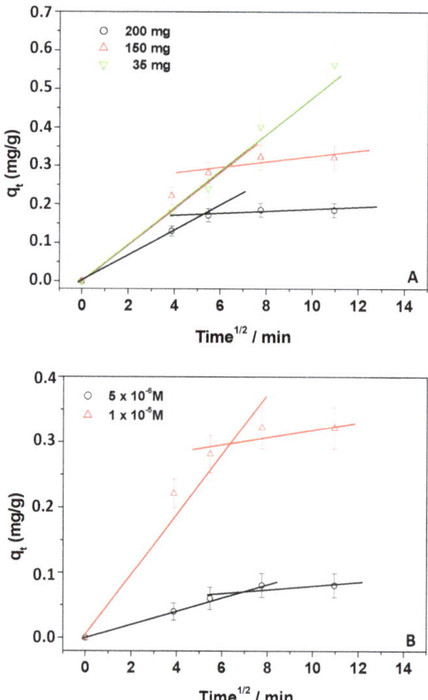

Figure 4. Weber–Morris plot applied to experiments of adsorption in which the amounts of the adsorbent (**A**) and Kp (**B**) were changed.

As a whole, the findings suggested that, under our experimental conditions, the Kp adsorption process could be described by two steps: (i) diffusion from the solution to the external surface of the adsorbent, and (ii) intra-particle adsorption and diffusion. Indeed, the experimental points could be divided into these two stages (Figure 4A,B). Moreover, during the second stage, since the ΔC of Kp decreased, the adsorption process decreased, reaching an equilibrium state. Once again, the exception was represented by the condition in which the smallest amount of chitosan was used. In this case, the regression line of the first stage passed through the origin of the plot, suggesting that the intra-particle diffusion was the rate-limiting step [31]. In this latter case, the number of available free sites present on the CH surface was lowered, and the diffusion controlled the process as previously supposed.

3.3. Effect of pH

To get insight into the nature of the adsorption process, the effect of pH during the adsorption of Kp onto CH was investigated, by adding either HCl or NaOH. To avoid changes in pH values during the adsorption process, due to the presence of acetic acid added during chitosan film preparation, the used adsorbent (150 mg) was neutralized with NaOH and washed several times with fresh water, until a neutral pH was achieved.

As reported in Figure 5A, the percentage of Kp adsorption was calculated at each pH value and contact time. The maximum adsorption occurred between pH 3 and pH 5, while it decreased after this pH value, with the lowest Kp removal at pH 12. Interestingly, at the beginning of the process, i.e., in the first 15 min, the Kp removal was approximately the same at pH 3 and 5, and it was reduced upon increasing the pH value. Instead, upon extending the contact time, the affinity at pH 3 was slightly reduced, and the results at pH 5 appeared the best. These findings were better evidenced by calculating the associated q_t values (Figure 5B).

Clearly, at both pH 3 and pH 5, the adsorption capacities were good, having the highest values, and, at pH 3, the $q_{t,max}$ was reached quickly in the first minute. In order to better understand this behavior, the zero-point charge (ZPC) of CH was determined using the drift method [14] (Figure 5C). The observed cross-section region of the curves in Figure 5C indicated that the pH_{ZPC} of CH was around pH 7. This means that, below pH 7, the chitosan amino groups were positively charged, while they were deprotonated toward the pH_{ZPC}. After this pH value, chitosan became mainly negatively charged [33]. The negative charge on the surface of the chitosan film in alkaline medium could be mainly ascribed to the presence of negative ions (OH^-) in solution, that would form a negative layer on the surface of chitosan. Moreover, in accordance with the carboxylic moieties present in the Kp chemical structure (see Figure 1A), the NSAID pK_a is reported to be around 4 [34]. This means that, below this pH value, Kp was present as a neutral molecule (Kp-H), while, above this value, it was present as an anionic one (Kp^-).

Thus, below pH 5, Kp was present as Kp-H, and CH showed a positively charged surface. As a result, a reduced affinity between Kp-H and the adsorbent was expected due to the reduced contribution of electrostatic interactions. However, since adsorption at pH 3 was quite significant, other forces should probably be considered during the process, such as dipole–dipole interactions and H-bonds. At pH 5, Kp was mainly present as Kp^- and, at the same time, chitosan was positively charged; thus, electrostatic interactions between the carboxylic moieties of Kp^- and the chitosan's positively charged amino groups took place, favoring pollutant removal. At pH > 5, i.e., pH 6, the chitosan amino groups were mainly deprotonated, thereby reducing the affinity between the adsorbent and the NSAID, thus suggesting an electrostatic repulsion between the negative Kp and CH charges [35–38]. This effect was more evident at pH 12, at which the adsorption was completely blocked. Figure 5D reports the possible scheme of interaction between Kp and CH, considering mainly electrostatic forces.

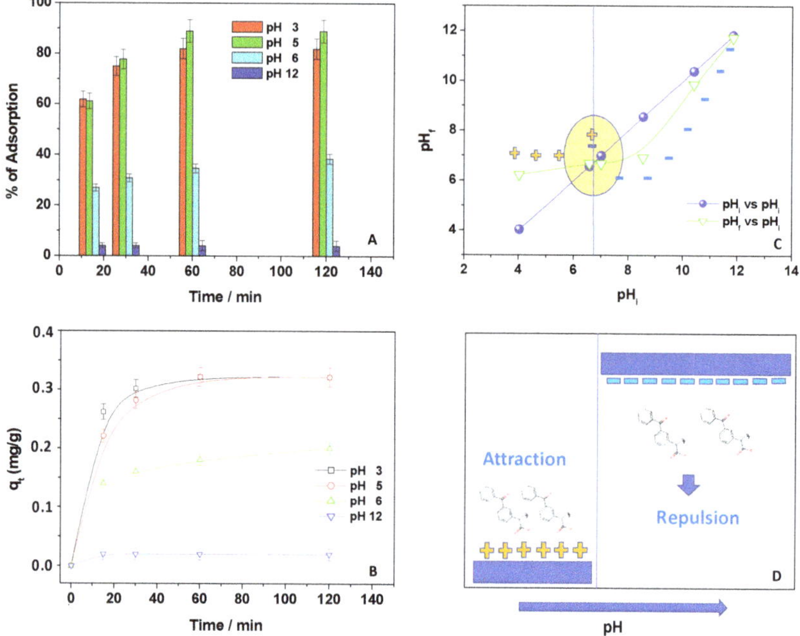

Figure 5. Percentage of Kp adsorption from an aqueous solution (1×10^{-5} M) when 150 mg of chitosan was used at different pH values (**A**); drift method to determine the zero-point charge of the adsorbent (**B**); adsorption capacities, q_t, referring to Kp adsorption from an aqueous solution (1×10^{-5} M) when 150 mg of chitosan was used at different pH values (**C**); cartoon depicting the interaction between Kp and the adsorbent (**D**).

3.4. Effect of Salts in Kp Solutions

With the aim of assessing the role of the electrostatic interaction between the Kp$^-$ anion and chitosan, some experiments were performed, changing the ionic strength of Kp solutions by adding electrolytes. By selecting NaCl as a model electrolyte at different concentrations, the experiments were performed using 150 mg of chitosan and 1×10^{-5} M Kp.

Thus, the Kp adsorption was evaluated, and the results obtained at 60 min of contact time are reported in Figure 6A. By changing the salt concentration from 0.01 M to 0.05 M, Kp removal decreased from 85%, in the absence of salt, to 20% with 0.05 M NaCl (Figure 6A). By choosing 0.01 M as the salt reference concentration, the electrolyte nature was changed. In particular, upon fixing the type of anion (Cl$^-$), the cation was changed by exploring the effects of Na$^+$, K$^+$, and Mg^{2+}. As reported in Figure 6B, upon decreasing the cation size from K$^+$ to Na$^+$, the Kp removal efficiency decreased and, upon changing the cation associated charge, using Mg^{2+}, the effect became more pronounced.

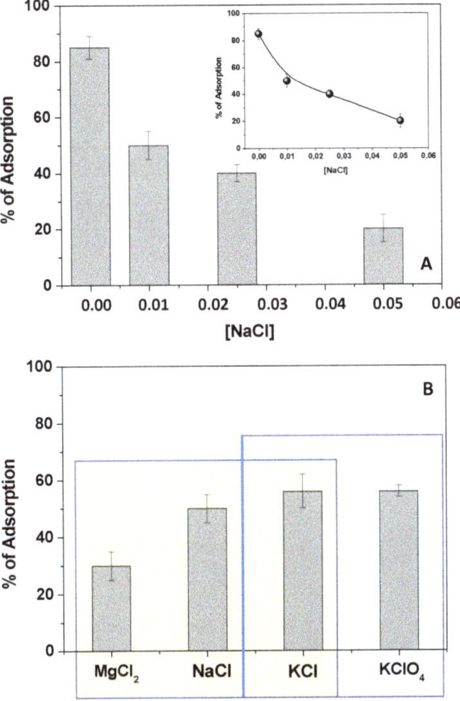

Figure 6. Percentage of Kp adsorption from an aqueous solution (1×10^{-5} M, pH 5) when 150 mg of chitosan was used at different concentrations of NaCl (**A**) and in the presence of different salts at 0.01 M (**B**).

These results suggested a cation-mediated shielding effect of the Kp negative charge, confirming the involvement of the Kp carboxylic moieties in its adsorption onto the chitosan film. Interestingly, by changing the type of anion, and choosing K^+ as the cation, the results reported in Figure 6B were obtained. The absence of significant changes in the removal of Kp indicated that the inorganic anions did not affect the adsorption. In fact, in general, if the anion affected the process, its effect would involve the shielding of CH positive charges onto the film surface, slowing down the Kp diffusion into the film, thus preventing adsorption. Instead, the cation effects occurred in solution, shielding the Kp negative charge, thereby reducing the adsorbate/adsorbent affinity.

3.5. Thermodynamic Analysis

The Kp adsorption process was investigated by adopting three temperature values, i.e., 278, 288, and 298 K, using 150 mg of chitosan and a 1×10^{-5} M Kp solution at pH 5. Figure 7 shows the obtained results in term of adsorption capacities (Figure 7A) and percentage of Kp adsorption (Figure 7B) at different contact times. Upon increasing the temperature values, the q_t values and the Kp adsorption percentage at the equilibrium increased, indicating the endothermic character of the process. With the aim of obtaining the thermodynamic parameters, the K_{eq} values were calculated at each temperature and, by using Equations (4) and (5), the correspondent $\Delta G°$ values were inferred (Table 2). The negative $\Delta G°$ values indicated the spontaneity of the Kp adsorption process onto chitosan. Furthermore, by plotting $\ln K_{eq}$ versus 1/T (Figure 7C) and applying Equation (6), $\Delta H°$ and $\Delta S°$ were also calculated (Table 2). In agreement with the literature [31], the positive values of $\Delta H°$

and $\Delta S°$ confirmed the endothermic character of the process and the increased randomness at the adsorbent–adsorbate interface, respectively.

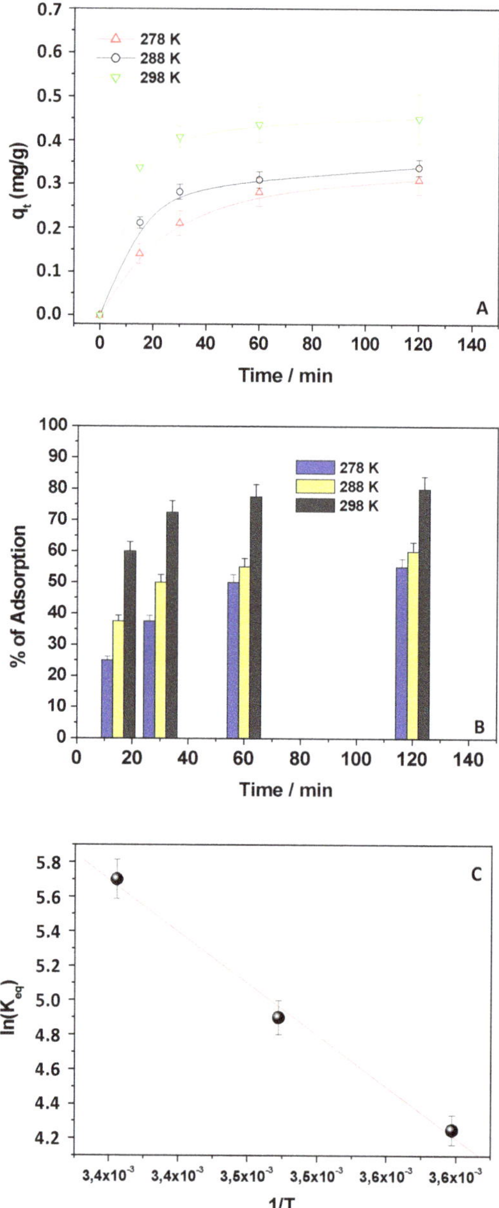

Figure 7. Adsorption capacities, q_t (**A**), and percentage of Kp adsorption (**B**) at different contact times, referring to a Kp aqueous solution (1×10^{-5} M, pH 5) when 150 mg of chitosan was used at different temperature values; plot of ln (K_{eq}) versus 1/T (**C**).

Table 2. Thermodynamic parameters.

T (K)	K_{eq}	Ln (K_{eq})	$\Delta H°$ (KJ/mol)	$\Delta S°$ (J/mol·K)	$\Delta G°_{278}$ (KJ/mol)
278	70	4.25			$-(9.7 \pm 7.8)$
288	135	4.90	$+(49.8 \pm 4.0)$	$+(214.2 \pm 14.0)$	$-(11.8 \pm 8.0)$
298	300	5.70			$-(14.0 \pm 8.2)$

3.6. Consecutive Cycles of Adsorption

With regard to a circular economy and sustainable application, the great performance of the proposed adsorbent was shown upon performing consecutive cycles of adsorption. By selecting 120 min as the contact time, a 1×10^{-5} M Kp solution at pH 5 was placed in contact with the adsorbent (150 mg) and, after almost complete adsorption (80% of Kp), the same film was used again to adsorb Kp from another solution at the same concentration. The Kp adsorption percentages reported in Table 3 were obtained. The results indicated that, after eight cycles of adsorption, the NSAID removal was still high, suggesting the possibility of using the same film several times for the removal of the NSAID, thus extending the adsorbent's lifetime.

Table 3. Consecutive cycles of adsorption referring to experiments of adsorption in which the Kp concentration was maintained at 1×10^{-5} M, pH 5, and the amount of chitosan was 150 mg.

N° of Cycles	Percentage of Adsorption
1	77 ± 5
2	82 ± 8
3	73 ± 7
4	77 ± 10
5	76 ± 5
6	73 ± 5
7	68 ± 8
8	68 ± 6

3.7. Release of Kp and Reuse of the Adsorbent

Another positive aspect of the proposed adsorbent was the possibility to recover the adsorbed pollutant, recycling both the molecule and the adsorbent itself. Indeed, due to the involvement of electrostatic interactions between Kp and chitosan, solutions containing salts were successfully used to obtain the desorption, thus reducing the environmental impact. Among the studied salts (see Figure 8), NaCl and $MgCl_2$ solutions were selected and compared. After Kp adsorption, with a contact time of 120 min, the chitosan film was placed in contact with aqueous solutions of NaCl and $MgCl_2$ at several concentrations, under continuous stirring. Once again, the Kp UV–Vis absorption spectrum, collected at 30 min, was used to monitor the Kp release. The percentage of the desorbed Kp, normalized for the adsorbed amount, was calculated, and the results are reported in Figure 8. The use of $MgCl_2$ ensured the total release of adsorbed Kp, upon using diluted solutions of salts (Figure 8A); on the other hand, the use of NaCl required a larger amount of salt. In fact, Kp release was only reasonable beyond a concentration of 0.25 M NaCl (Figure 8B).

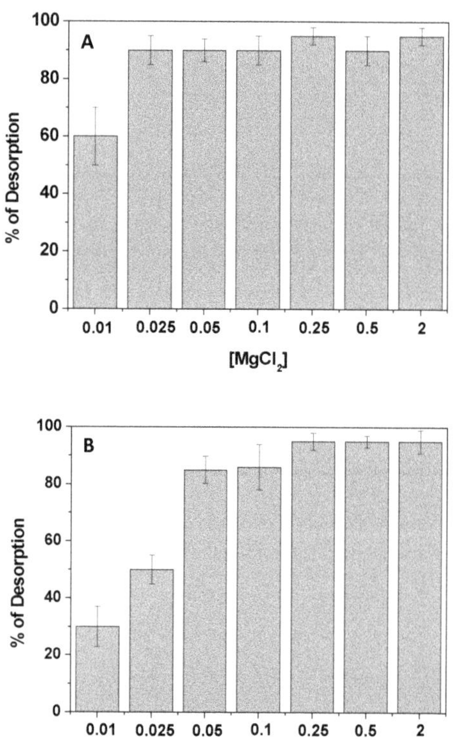

Figure 8. Percentage of Kp desorption, calculated by adopting 30 min as the contact time, in the presence of different concentrations of MgCl$_2$ (**A**) and NaCl (**B**).

However, it is worth pointing out that the use of MgCl$_2$ induced the slight degradation of the chitosan film. Thus, with a view of a more sustainable procedure for cleaner production and pollution prevention, the use of NaCl 0.25 M is suggested for Kp recovery, applying 30 min as the contact time for its desorption. Several cycles of adsorption/desorption were performed, and the percentage of the adsorbed/desorbed Kp for each cycle is reported in Table 4. Once again, the desorbed Kp percentage was normalized with respect to the correspondent adsorbed Kp. After each cycle, the adsorbent was washed with fresh water, and subsequently placed in contact with a fresh Kp solution. Interestingly, despite the same film being used several times, the efficiencies of Kp removal and recovery were not affected, and the obtained results suggested the possibility of reducing the procedure-associated costs and the amount of secondary pollutants potentially released into the environment, suggesting not only possible Kp reuse, but also adsorbent recycling.

Table 4. Consecutive cycles of adsorption and desorption (obtained in the presence of 0.25 M NaCl, with a contact time of 30 min) referring to experiments of adsorption in which the Kp concentration was maintained at 1×10^{-5} M, pH 5, and the amount of chitosan was 150 mg.

N° of Cycles	Percentage of Adsorption	Percentage of Desorption
1	84 ± 5	95 ± 3
2	84 ± 5	92 ± 7
3	89 ± 5	95 ± 5
4	90 ± 5	95 ± 5
5	89 ± 5	95 ± 5
6	91 ± 5	95 ± 5
7	89 ± 4	95 ± 5
8	85 ± 4	93 ± 6
9	80 ± 4	90 ± 5
10	75 ± 4	90 ± 4

4. Conclusions

Due to concerns related to water decontamination from pollutants, this paper reported the use of chitosan films for the removal of Kp from water, with regard to more ustainable and greener industrial applications. The bio-sorption process is very interesting, since about 90% of the Kp was removed in 2 h at most. The sorption process was investigated at several temperature values, indicating that, upon increasing the temperature, the removal of Kp from water increased. The thermodynamic parameters were calculated, observing that the process was spontaneous ($\Delta G° < 0$) and endothermic ($\Delta H° < 0$), and that it occurred with an increase in entropy. From a kinetic point of view, the pseudo second-order kinetic model agreed well with the experimental data, indicating that the bio-sorption was dependent on the amounts of Kp and adsorbent. In fact, upon increasing the amount of CH and the concentration of the Kp solution, the pollutant removal efficiencies affected the adsorption capacities. The high affinity between the NSAID and CH was ascribed to the presence of electrostatic interactions. Indeed, the adsorption was largely affected by the pH value of the Kp solution and by the presence of salts.

Considering that the pK_a of Kp is about 4, the maximum removal was observed at pH 5, while, upon further increasing the pH values of the Kp solution, the NSAID removal decreased. In accordance with the measured pH_{ZPC} of the CH (~7), below this pH value, the interaction between Kp^- and CH was observed. However, below the pK_a of the Kp, the main form Kp-H reduced the affinity toward the adsorbent. The same finding was observed at basic pH values, conditions in which the adsorbent was negatively charged and repulsions between Kp^- and the adsorbent occurred. An interaction between the Kp carboxylic moiety and the chitosan amino groups was, thus, proposed and confirmed by experiments of adsorption performed in the presence of salts, which inhibited the Kp removal. Interestingly, the use of 0.25 M NaCl was found to be suitable for the desorption of the adsorbed Kp, enabling the reuse of the pollutant and the recycling of the adsorbent for several cycles, extending the CH lifetime. The reuse of the recovered pollutants and the use of CH, proposed in this work, exhibit great benefits for the environment through the reuse of products for cleaner production technologies.

Author Contributions: Conceptualization, V.R.; methodology, J.G., R.R., and P.F.; software, V.R. and S.N.; validation, V.R. and P.C.; formal analysis, J.G. and V.R.; investigation, V.R. and P.F.; resources, P.C.; data curation, V.R. and P.C.; writing—original draft preparation, V.R.; visualization, V.R. and P.F.; supervision, P.C.; project administration, P.C. and P.F.; funding acquisition, P.C. and P.F.

Funding: This research was funded by the LIFE+ European Project named LIFE CLEAN UP ("Validation of adsorbent materials and advanced oxidation techniques to remove emerging pollutants in treated wastewater"—LIFE 16 ENV/ES/000169).

Acknowledgments: We gratefully acknowledge Sergio Nuzzo for skillful and technical assistance.

Conflicts of Interest: The authors declare no conflicts of interest.

References

1. Divya, K.; Rebello, S.; Jisha, M.S. A Simple and Effective Method for Extraction of High Purity Chitosan from Shrimp Shell Waste. In Proceedings of the International Conference on Advances in Applied Science and Environmental Engineering—ASEE 2014, Kuala Lumpur, Malaysia, 2–3 August 2014. [CrossRef]

2. Teli, M.D.; Sheikh, J. Extraction of chitosan from shrimp shells waste and application in antibacterial finishing of bamboo rayon. *Int. J. Biol. Macromol.* **2012**, *50*, 1195–1200. [CrossRef] [PubMed]

3. Ahing, F.A.; Wid, N. Extraction and Characterization of Chitosan from Shrimp Shell Waste in Sabah. *Trans. Sci. Technol.* **2016**, *3*, 227–237.

4. Premasudha, P.; Vanathi, P.; Abirami, M. Extraction and Characterization of Chitosan from Crustacean Waste: A Constructive Waste Management Approach. *Int. J. Sci. Res.* **2017**, *6*, 1194–1198.

5. Yong, S.K.; Shrivastava, M.; Srivastava, P.; Kunhikrishnan, A.; Bolan, N. Environmental applications of chitosan and its derivatives. *Rev. Environ. Contam. Toxicol.* **2015**, *233*, 1–43. [CrossRef] [PubMed]

6. Kanmani, P.; Aravind, J.; Kamaraj, M.; Sureshbabu, P.; Karthikeyan, S. Environmental applications of chitosan and cellulosic biopolymers: A comprehensive outlook. *Bioresour. Technol.* **2017**, *242*, 295–303. [CrossRef]

7. Rizzi, V.; Longo, A.; Placido, T.; Fini, P.; Gubitosa, J.; Sibillano, T.; Giannini, C.; Semeraro, P.; Franco, E.; Ferrandiz, M.; et al. A comprehensive investigation of chitosan/dyes blended films for green chemistry applications. *J. Appl. Polym. Sci.* **2017**, *135*, 45945. [CrossRef]

8. Semeraro, P.; Fini, P.; D'Addabbo, M.; Rizzi, V.; Cosma, P. Removal from wastewater and recycling of azo textile dyes by alginate-chitosan beads. *Int. J. Environ. Agric. Biotechnol. (IJEAB)* **2017**, *2*, 1835–1850. [CrossRef]

9. Rizzi, V.; Longo, A.; Fini, P.; Semeraro, P.; Cosma, P.; Franco, E.; García, R.; Ferrándiz, M.; Núñez, E.; Gabaldón, J.A.; et al. Applicative Study (Part I): The Excellent Conditions to Remove in Batch Direct Textile Dyes (Direct Red, Direct Blue and Direct Yellow) from Aqueous Solutions by Adsorption Processes on Low-Cost Chitosan Films under Different Conditions. *Adv. Chem. Eng. Sci.* **2014**, *4*, 454–469. [CrossRef]

10. Kandile, N.G.; Mohamed, H.M. Chitosan nanoparticle hydrogel based sebacoyl moiety with remarkable capability for metal ion removal from aqueous systems. *Int. J. Biol. Macromol.* **2019**, *122*, 578–586. [CrossRef]

11. Rizzi, V.; Fini, P.; Fanelli, F.; Placido, T.; Semeraro, P.; Sibillano, T.; Fraix, A.; Sortino, S.; Agostiano, A.; Giannini, C.; et al. Molecular interactions, characterization and photoactivity of Chlorophyll a/Chitosan/2-HP-β-Cyclodextrin composite films as functional and active surfaces for ROS production. *Food Hydrocoll.* **2016**, *58*, 98–112. [CrossRef]

12. Rizzi, V.; Fini, P.; Semeraro, P.; Cosma, P. Detailed Investigation of ROS arisen from Chlorophyll a/Chitosan based-biofilm. *Colloids Surf. B* **2016**, *142*, 239–247. [CrossRef] [PubMed]

13. Rizzi, V.; Prasetyanto, E.A.; Chen, P.; Gubitosa, J.; Fini, P.; Agostiano, A.; De Cola, L.; Cosma, P. Amino grafted MCM-41 as highly efficient and reversible ecofriendly adsorbent material for the Direct Blue removal from wastewater. *J. Mol. Liq.* **2019**, *273*, 435–446. [CrossRef]

14. Rizzi, V.; Fiorini, F.; Lamanna, G.; Gubitosa, J.; Prasetyanto, E.A.; Fini, P.; Fanelli, F.; Nacci, A.; De Cola, L.; Cosma, P. Polyamidoamine-Based Hydrogel for Removal of Blue and Red Dyes from Wastewater. *Adv. Sustain. Syst.* **2018**, *2*, 1700146. [CrossRef]

15. Gomes, A.R.; Justino, C.; Rocha-Santos, T.; Freitas, A.C.; Duarte, A.C.; Pereira, R. Review of the ecotoxicological effects of emerging contaminants to soil biota. *J. Environ. Sci. Health A Environ. Sci. Eng.* **2017**, *52*, 992–1007. [CrossRef]

16. Rizzi, V.; D'Agostino, F.; Fini, P.; Semeraro, P.; Cosma, P. An interesting environmental friendly cleanup: The excellent potential of olive pomace for disperse blue adsorption/desorption from wastewater. *Dyes Pigments* **2017**, *140*, 480–490. [CrossRef]

17. Jeirani, Z.; Niu, C.H.; Soltan, J. Adsorption of emerging pollutants on activated carbon. *Rev. Chem. Eng.* **2017**, *33*, 491–522. [CrossRef]

18. Rizzi, V.; Lacalamita, D.; Gubitosa, J.; Fini, P.; Petrella, A.; Romita, R.; Agostiano, A.; Gabaldón, J.A.; Fortea Gorbe, M.I.; Gómez-Morte, T.; et al. Removal of tetracycline from polluted water by chitosan-olive pomace adsorbing films. *Sci. Total Environ.* **2019**, *693*, 133620. [CrossRef]

19. Rizzi, V.; Mongiovì, C.; Fini, P.; Petrella, A.; Semeraro, P.; Cosma, P. Operational parameters affecting the removal and recycling of direct blue industrial dye from wastewater using bleached oil mill waste as alternative adsorbent material. *Int. J. Environ. Agric. Biotechnol. (IJEAB)* **2017**, *2*, 1560–1572. [CrossRef]

20. Dai, Y.; Zhang, K.; Meng, X.; Li, J.; Guan, X.; Sun, Q.; Sun, Y.; Wang, W.; Lin, M.; Liu, M.; et al. New use for spent coffee ground as an adsorbent for tetracycline removal in water. *Chemosphere* **2019**, *215*, 163–172. [CrossRef]

21. Cuklev, F.; Fick, J.; Cvijovic, M.; Kristiansson, E.; Förlin, L.; Larsson, D.G.J. Does ketoprofen or diclofenac pose the lowest risk to fish? *J. Hazard. Mater.* **2012**, *229–230*, 100–106. [CrossRef]

22. Diniz, M.S.; Salgado, R.; Pereira, V.J.; Carvalho, G.; Oehmen, A.; Reis, M.A.M.; Noronha, J.P. Ecotoxicity of ketoprofen, diclofenac, atenolol and their photolysis byproducts in zebrafish (Danio rerio). *Sci. Total Environ.* **2015**, *505*, 282–289. [CrossRef] [PubMed]

23. Hasan, H.A.; Abdullah, S.R.S.; Al-Attabi, A.W.N.; Nash, D.A.H.; Anuar, N.; Rahman, N.A.; Titah, H.S. Removal of ibuprofen, ketoprofen, COD and nitrogen compounds from pharmaceutical wastewater using aerobic suspension-sequencing batch reactor (ASSBR). *Sep. Purif. Technol.* **2016**, *157*, 215–221. [CrossRef]

24. Jankunaite, D.; Tichonovas, M.; Buivydiene, D.; Radziuniene, I.; Racys, V.; Krugly, E. Removal of Diclofenac, Ketoprofen, and Carbamazepine from Simulated Drinking Water by Advanced Oxidation in a Model Reactor. *Water Air Soil Pollut.* **2017**, *228*, 1–15. [CrossRef]

25. Nagy, Z.M.; Molnár, M.; Fekete-Kertész, I.; Molnár-Perl, I.; Fenyvesi, É.; Gruiz, K. Removal of emerging micropollutants from water using cyclodextrin. *Sci. Total Environ.* **2014**, *485–486*, 711–719. [CrossRef]

26. Madikizela, L.M.; Zunngu, S.S.; Mlunguza, N.Y.; Tavengwa, N.T.; Mdluli, P.S.; Chimuka, L. Application of molecularly imprinted polymer designed for the selective extraction of ketoprofen from wastewater. *Water SA* **2018**, *44*, 406–418. [CrossRef]

27. Rizzi, V.; Romanazzi, F.; Gubitosa, J.; Fini, P.; Romita, R.; Agostiano, A.; Petrella, A.; Cosma, P. Chitosan Film as Eco-Friendly and Recyclable Bio-Adsorbent to Remove/Recover Diclofenac, Ketoprofen, and their Mixture from Wastewater. *Biomolecules* **2019**, *9*, 571. [CrossRef]

28. Sekulic, M.T.; Boskovic, N.; Milanovic, M.; Letic, N.G.; Gligoric, E.; Papa, S. An insight into the adsorption of three emerging pharmaceutical contaminants on multifunctional carbonous adsorbent: Mechanisms, modelling and metal coadsorption. *J. Mol. Liq.* **2019**, *284*, 372–382. [CrossRef]

29. Anbia, M.; Salehi, S. Removal of acid dyes from aqueous media by adsorption onto amino-functionalized nanoporous silica SBA-3. *Dyes Pigments* **2012**, *94*, 1–9. [CrossRef]

30. Rizzi, V.; D'Agostino, F.; Gubitosa, J.; Fini, P.; Petrella, A.; Agostiano, A.; Semeraro, P.; Cosma, P. An Alternative Use of Olive Pomace as a Wide-Ranging Bioremediation Strategy to Adsorb and Recover Disperse Orange and Disperse Red Industrial Dyes from Wastewater. *Separations* **2017**, *4*, 29. [CrossRef]

31. Antunes, M.; Esteves, V.I.; Guégan, R.; Crespo, J.S.; Fernandes, A.N.; Giovanela, M. Removal of diclofenac sodium from aqueous solution by Isabel grape bagasse. *Chem. Eng. J.* **2012**, *192*, 114–121. [CrossRef]

32. Lin, K.Y.A.; Yang, H.; Lee, W.D. Enhanced removal of diclofenac from water using a zeolitic imidazole framework functionalized with cetyltrimethylammonium bromide (CTAB). *RSC Adv.* **2015**, *5*, 81330–81340. [CrossRef]

33. Chang, J.H.; Chen, C.L.; Ellis, A.V.; Tung, C.H. Studies of Chitosan at Different pH's in the Removal of Common Chlorinated Organics from Wastewater. *Int. J. Appl. Sci. Eng.* **2012**, *10*, 293–306.

34. Meloun, M.; Bordovská, S.; Galla, L. The thermodynamic dissociation constants of four non-steroidal anti-inflammatory drugs by the least-squares nonlinear regression of multiwavelength spectrophotometric pH-titration data. *J. Pharm. Biomed. Anal.* **2007**, *45*, 552–564. [CrossRef] [PubMed]

35. Desbrières, J.; Guibal, E. Chitosan for wastewater treatment. *Polym. Int.* **2018**, *67*, 7–14. [CrossRef]

36. Pereira, K.A.A.; Osório, L.R.; Silva, M.P.; Sousa, K.S.; Da Silva Filho, E.C. Chemical Modification of Chitosan in the Absence of Solvent for Diclofenac Sodium Removal: pH and Kinetics Studies. *Mater. Res.* **2014**, *17*, 141–145. [CrossRef]

37. Zhang, Y.; Shen, Z.; Dai, C.; Zhou, X. Removal of selected pharmaceuticals from aqueous solution using magnetic chitosan: Sorption behavior and mechanism. *Environ. Sci. Pollut. Res.* **2014**, *21*, 12780–12789. [CrossRef] [PubMed]
38. Carvalho, T.O.; Matias, A.E.B.; Braga, L.R.; Evangelista, S.M.; Prado, A.G.S. Calorimetric studies of removal of nonsteroidal anti-inflammatory drugs diclofenac and dipyrone from water. *J. Therm. Anal. Calorim.* **2011**, *106*, 475–481. [CrossRef]

Article

Rendering Mortars Reinforced with Natural Sheep's Wool Fibers

Cinthia Maia Pederneiras [1,2], Rosário Veiga [2] and Jorge de Brito [1,*]

[1]   CERIS, Instituto Superior Técnico, University of Lisbon, 1049-001 Lisbon, Portugal;
       cinthiamaia@tecnico.ulisboa.pt
[2]   National Laboratory for Civil Engineering, 1700-066 Lisbon, Portugal; rveiga@lnec.pt
*   Correspondence: jb@civil.ist.utl.pt; Tel.: +351-218-418-118

Received: 12 October 2019; Accepted: 4 November 2019; Published: 6 November 2019

Abstract: The susceptibility of rendering mortars to cracking is a complex phenomenon. Fibers have been incorporated in mortars to ensure the durability of the render and can improve the flexural strength, fracture toughness, and impact resistance of the mortars. Aside from the better cracking performance of fiber reinforced mortars, natural fibers have been a path to reducing the environmental impacts of construction materials. Recycling has high sustainability-related potential as it can both mitigate the amount of waste being inadequately disposed and reduce the consumption of natural raw materials. Studies on the incorporation of waste in civil engineering materials have been growing, and recycled fibers may be feasible to incorporate in mortars. Natural fibers are considered as a viable replacement for synthetic ones. Several studies have investigated vegetal fibers in cementitious composites. However, only a few have focused on the incorporation of waste animal-based fiber. The aim of this work is to analyze the feasibility of the use of natural sheep's wool fibers on the reinforcement of mortars and in particular to improve their cracking behavior. For this purpose, two different binders were used: cement and cement-lime mortars were produced. The incorporation of 10% and 20% (in volume) of 1.5 cm and 3.0 cm wool fibers was analyzed. The results show that the incorporation of wool fibers increased the ductility of the mortars and improved their mechanical properties.

Keywords: render; cement and cement-lime reinforced mortars; natural fiber; sheep's wool; sustainability

---

1. Introduction

The construction sector has been trying to reduce its environmental impacts. Eco-friendly constituents have been an alternative to develop new cementitious materials. In order to enhance the cracking performance of mortars, the incorporation of natural fibers can contribute to better performance while ensuring a sustainable approach. According to previous studies, natural fibers such as sheep wool may improve the ductility of cementitious composites and also provide an adequate disposal of the waste [1].

The incorporation of fibers enhances a better post-cracking behavior due to the higher fracture toughness, flexural strength, and impact resistance of the mortars [2]. The benefits of fiber reinforcement in cementitious materials depend on the fiber type, their geometry, and their volume ratio and distribution [3]. The use of natural fibers compared to man-made fibers has been achieving environmental, energy, and resource conservation benefits [4]. There are three types of natural fibers: plant-based, mineral-derived, and animal-based. In this paper, the study focused on animal-based fibers, namely sheep's wool fibers.

Natural sheep wool is considered as waste on a large-scale, taking into account that 270.000 tons of wool are produced by 90 million sheep in Europe [3]. Indeed, 75% of the wool produced (around

150 million tons per year) is rejected by the textile industry [4,5]. Most of this material does not have a proper disposal method. Furthermore, the wool fibers have an elastic modulus of about 1–4 GPa, which can be comparable to the modulus of plastic fibers [5].

Several studies have been carried out to investigate the incorporation of wool fibers in cementitious materials to improve the thermal insulation properties [6–9]. However, only a few studies, described next, have incorporated wool fibers with the purpose of improving the mechanical performance of cementitious composites.

Alyousef et al. [10] analyzed the mechanical properties of reinforced concrete with wool fibers and found that the incorporation of fibers decreased the workability of concrete. Concerning the mechanical behavior, the fibers enhanced the ductility and flexural capacity of the composite. In terms of compressive strength, the incorporation of fibers implied a reduction of strength. This reduction was explained by the authors by the incorporation of wool up to 6% by weight of cement in the concrete, which led to a reduction of the total binder in the mix.

Fantilli et al. [5] investigated the incorporation of 1% (in volume) of wool fibers in cementitious reinforced-mortars. The authors observed that the fibers reduced the brittle nature of the mortars due to the development of bridge mechanisms between the crack borders. In this research, an improvement in the fracture toughness in the reinforced mortars of 300% relative to the control mortar is reported. It could also be seen that the wool fibers promoted a reduction in the plastic shrinkage. The cementitious mortars reinforced with wool fibers presented the same mechanical performance (i.e., strengths and modulus of elasticity) as those in mortars with the most common vegetal fibers.

Giosué et al. [11] investigated the replacement of 25% (in volume) of aggregates with wool fibers in lime-based lightweight mortars. The modified mortars were tested in the fresh and hardened state concerning workability, mechanical strength, and hygro-thermal properties. The results showed an increase of about 30% in the flexural strength of the modified mortars.

Kesikidou and Stefanidou [12] investigated the incorporation of natural fibers in mortars. The authors analyzed mortars with two different binders, incorporating vegetal fibers such as jute, coconut, and kelp and showed that the natural fibers performed differently in relation to cement or lime-based binder. Lime-reinforced mortars presented a higher increase in flexural and compressive strength when compared to the cement-based reinforced mortars. Therefore, the compatibility of the fibers with the mortar's composition should be evaluated.

As a conclusion of this review, it was found that, aside from all of the studies carried out on mortars with natural fibers, the incorporation of wool fibers in rendering cementitious mortars could not be found in the technical literature. This is relevant because rendering mortars mainly have the function of protecting the substrate. This means that a high compressive strength is not often necessary, but is important to minimize cracking. As a consequence, usually low strength mortars with a low binder/aggregate ratio are used, and from this aspect, leads to a different microstructure and a poorer adhesion between the binder, aggregates, and fibers.

Additionally, the compositions used result in a low modulus of elasticity, possibly nearer the modulus of elasticity of natural fibers than in the case of structural mortars. Finally, for rendering mortars, as a favorable cracking behavior is more important than a high mechanical strength, characteristics such as a low modulus of elasticity and ductility are mainly required. Therefore, the novelty of this work was in analyzing the feasibility of the incorporation of wool fibers in rendering mortars and their efficacy in improving those specific properties. Cement and cement-lime mortars with low binder/aggregate volumetric ratios of 1:4 and 1:3 were produced, respectively, with different fibers lengths of 1.5 and 3.0 cm, respectively. The ratios of incorporation were 10% and 20% of the total mortar volume.

The mortars' properties considered relevant for rendering mortars, namely workability, bulk density, dynamic modulus of elasticity, ultra-sound pulse velocity, flexural and compressive strengths, and protection to water action, were evaluated in the fresh and hardened states.

From the results, it could be seen that the incorporation of wool fibers in rendering mortars with a volumetric ratio of 1:4 (cement: aggregates) and 1:1:6 (cement:air-lime:aggregates) improved the cracking behavior of the materials as the modified mortars, in general, presented a higher flexural strength and a lower modulus of elasticity when compared to the plain mortars.

2. Experimental Program

The aim of this research was to evaluate the influence of natural fiber waste, namely sheep wool, incorporated in rendering mortars.

Cement mortars and cement-air lime-based mortars were produced. The volumetric proportions were 1:4 (cement: aggregates) and 1:1:6 (cement:air-lime:aggregates). Aside from the binder, the length and volume ratio were parameters used to formulate the analyzed mortars.

Mortars were identified as follows:

- REF 1:4 (0% of incorporation—reference cement mortar)
- W 1.5_10%c (10% of incorporation of 1.5 cm long wool fibers—cement mortar)
- W 3.0_10%c (10% of incorporation of 3.0 cm long wool fibers—cement mortar)
- W 1.5_20%c (20% of incorporation of 1.5 cm long wool fibers—cement mortar)
- W 3.0_20%c (20% of incorporation of 3.0 cm long wool fibers—cement mortar)
- REF 1:1:6 (0% of incorporation—reference cement-lime mortar)
- W 1.5_10%cl (10% of incorporation of 1.5 cm long wool fibers—cement-lime mortar)
- W 3.0_10%cl (10% of incorporation of 3.0 cm long wool fibers—cement-lime mortar)
- W 1.5_20%cl (20% of incorporation of 1.5 cm long wool fibers—cement-lime mortar)
- W 3.0_20%cl (20% of incorporation of 3.0 cm long wool fibers—cement-lime mortar)

2.1. Materials

The materials used were cement, air-lime, sand, and wool fibers. The binder used on the cement mortars was CEM II/B-L 32.5 N, according to EN 197-1 [13]. The calcium hydrated lime powder—air lime—used was class CL80-S, according to EN 459-1 [14]. The natural silica sand was previously washed, calibrated, and sieved for a required size distribution. The wool fibers were washed with neutral detergent, and dried at 40 °C. This procedure was applied to remove the impurities. The fiber length was obtained by manually cutting the waste material. The fibers were added in order to ensure a homogeneous dispersion in each mix composition. The homogeneity of the fibers in the mix was implemented by distributing them in a properly closed receptacle and blowing compressed air over them in order to achieve an adequate dispersion before adding to the mix. Figure 1 presents the wool fibers used in the experimental campaign. The apparent bulk density of the constituents of the mortars produced is presented in Table 1.

2.2. Methods

All tests carried out in the experimental program are described in Table 3.

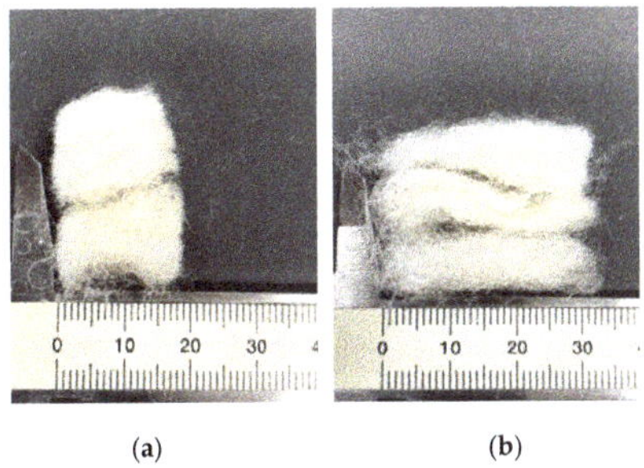

(a)   (b)

Figure 1. Sheep wool fiber used: (**a**) 1.5 cm long (**b**) 3.0 cm long.

Table 1. Apparent bulk density of the constituents.

Component	Apparent Bulk Density (kg/m^3)
Cement	975.5
Air-lime	565.7
Sand	1230.8
Wool 1.5 cm	4.25
Wool 3.0 cm	2.53

Table 2 presents the composition of the mortars produced in this work.

Table 2. Composition of the mortar mixes by mass.

Mortar	Cement (g)	Air-lime (g)	Sand (g)	Water (g)	Fibers (g)
REF 1:4	487.8	-	2461.6	445	-
W 1.5_10%c	439.1	-	2215.4	405	1.1
W 3.0_10%c	439.1	-	2215.4	435	0.6
W 1.5_20%c	390.2	-	1969.3	350	2.1
W 3.0_20%c	390.2	-	1969.3	350	1.3
REF 1:1:6	304.8	176.8	2307.8	465	-
W 1.5_10%cl	274.4	159.1	2077.0	435	1.1
W 3.0_10%cl	274.4	159.1	2077.0	420	0.6
W 1.5_20%cl	243.9	141.4	1846.2	370	2.1
W 3.0_20%cl	243.9	141.4	1846.2	370	1.3

Table 3. Experimental campaign tests.

Test	European Standard	Samples	Specimens	Age (days)
Apparent bulk density	Cahier 2669-4 [15]	6	Cement, lime, sand and fibers	-
Consistence by flow table	EN 1015-3 [16]	3	Fresh mortar	-
Bulk density	EN 1015-6 [17]	3	Fresh mortar	-
Dry bulk density	EN 1015-10 [18]	3	Hardened mortar	28, 90, and 180
Dynamic modulus of elasticity by resonance frequency method	EN 14146 [19]	3	Hardened mortar	28, 90. and 180
Ultra-sound pulse velocity	EN 12504-4 [20]	1	Hardened mortar	28, 90, and 180
Flexural and compressive strengths	EN 1015-11 [21]	3	Hardened mortar	28
Open porosity	EN 1936 [22]	3	Hardened mortar	28

For the hardened mortar tests, prismatic samples (40 mm × 40 mm × 160 mm) were used, in accordance with European Standards.

For ultra-sound pulse velocity, the direct and indirect methods were used for the measurements. In the direct transmission method, the electrodes are placed on the opposite surfaces of the specimen. In the indirect method, the electrodes are positioned on the same surface of the prism: the transmitter electrode is fixed at a specific point and the receptor moves over the specimen, and at different distances, the transmission time is measured allowing for the velocity to be calculated.

In order to analyze the susceptibility to cracking of the mortars produced in this work, some parameters were calculated. The Center Scientifique et Technique du Bâtiment (CSTB) [15] refers to the dynamic modulus of elasticity and flexural strength ratio (E/σ_f) as indicators of the mortar's ability to resist cracking. This criterion is based on the fact that a lower dynamic modulus of elasticity provides a higher deformation capacity of the material, and a greater flexural strength induces the material to withstand tensions without cracking. Therefore, the tendency to crack due to restrained shrinkage is greater when the ratio between the modulus of elasticity and tensile strength is high.

Another parameter to evaluate the susceptibility to crack is based on the flexural and compressive strengths. The ductility of the material can be associated to this ratio (σ_f/σ_c) (i.e., the mortar is considered more ductile when this value is closer to 1). Ductility is a measure of the deformability of the material before fracture. Cracking resistance is correlated with the deformation capacity of the mortar and its ability to absorb stress without cracking [23].

The ability to absorb energy before fracture is correlated with the toughness of the mortar. The fracture toughness was calculated by the total area under the strain–stress curve of the results of flexural strength at 28 days.

Regardless of the binder used, all hardened mortars were cured as specified by EN 1015-11 [21]. The specimens were kept in molds for two days at a temperature of 20 ± 2 °C and a relative humidity of 95 ± 5%. After demolding, all specimens were maintained in the same conditions for a total of seven days. After that, the specimens were kept at 20 ± 2 °C and the relative humidity was reduced to 65 ± 5%, until testing.

3. Results and Analysis

3.1. Fresh State

The mortar's workability was measured by the consistency test. To improve the comparability of the results, the values were limited to 140 ± 5 mm. The mortars presented a stiff consistence, but an application on a brick was carried out to ensure that an adequate workability was achieved, as shown in Figure 2.

Figure 2. Mortar application on a brick.

It was noticed that the fibers kept the mortar agglutinated. Although the workability was acceptable, the flor value did not increase due to the fibers' agglutinating action. In order to illustrate this behavior, Figure 3 presents the flow table test of the W 1.5_10%c sample.

Figure 3. Flow table test for the modified mortar.

Table 4 presents the results of the fresh mortars' properties.

Table 4. Fresh mortar properties.

Mortar	w/b Ratio	Consistency (mm)	Bulk Density (kg/m^3)
REF 1:4	0.91	140	2005.3
W 1.5_10%c	0.92	141	1978.1
W 3.0_10%c	0.98	139	1969.5
W 1.5_20%c	0.89	135	1936.5
W 3.0_20%c	0.89	141	2002.9
REF 1:1:6	0.98	140	1998.8
W 1.5_10%cl	1.02	141	1980.3
W 3.0_10%cl	0.98	141	1977.8
W 1.5_20%cl	0.97	139	1970.3
W 3.0_20%cl	0.97	139	1985.9

In all cement-lime mortars, the water/binder ratio was higher than that of the cement mortars. The incorporation of 10% of wool fibers increased the amount of water needed to maintain the workability. This could be due to the morphology of the fiber, which is composed of keratin filaments [5]. However, the incorporation of 20% of fibers again decreased the water/binder ratio to values similar to the control mortar. This trend reversion may be due to the fact that the longer fibers have a lower bulk density and thus a lower weight of incorporated fibers is actually used (Table 2). The modified mortars presented a lower bulk density than that of the reference mortars.

3.2. Hardened State

3.2.1. Dry Bulk Density of the Hardened Mortars

The dry bulk density of the hardened mortars was determined at 28, 90, and 180 days and the results are presented in Figure 4. The same trend as for the fresh state was noticed: the incorporation of wool fibers reduced the dry bulk density of the mortars due to the low bulk density of the fibers. Giosué et al. [11] found similar results (i.e., the bulk density of the reinforced mortars had a decrease of about 13% compared to the control mortar at 28 days).

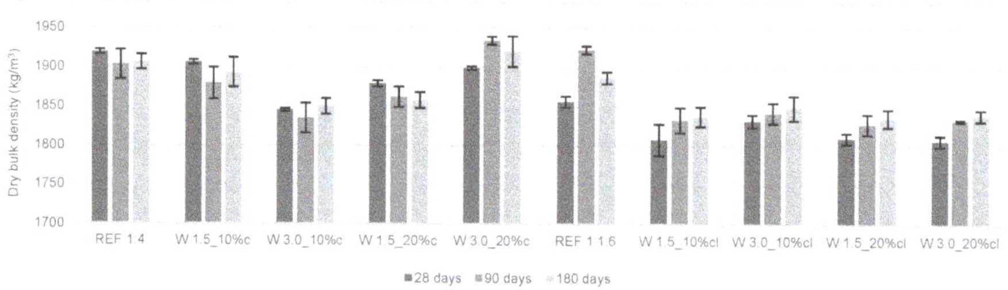

Figure 4. Dry bulk density of the hardened mortars.

In general, the dry bulk density of the cement mortars decreased from 28 to 90 days. Dry bulk density is defined as the ratio between mass and volume. Thus, the variations in weight and volume during time explain the variations in dry bulk density. In a previous study, a similar trend of decrease from 28 to 90 days was found [24], possibly because the mass reduction overlapped the volume reduction.

The opposite effect occurred in the cement-lime mortars, which presented an increase in this property over time. In fact, cement-lime mortars also have shrinkage with the consequent reduction of volume, and possibly have an increase in weight due to a higher carbonation reaction (by comparison with cement-only mortars). Previous works have shown this same trend for air lime mortars [25].

3.2.2. Dynamic Modulus of Elasticity of the Hardened Mortars

The modulus of elasticity measures the ability of the rendering mortars to absorb deformations. The renders should be able to withstand higher internal stresses without cracking. The dynamic modulus of elasticity was determined at 28, 90, and 180 days. The results are presented in Figures 5 and 6.

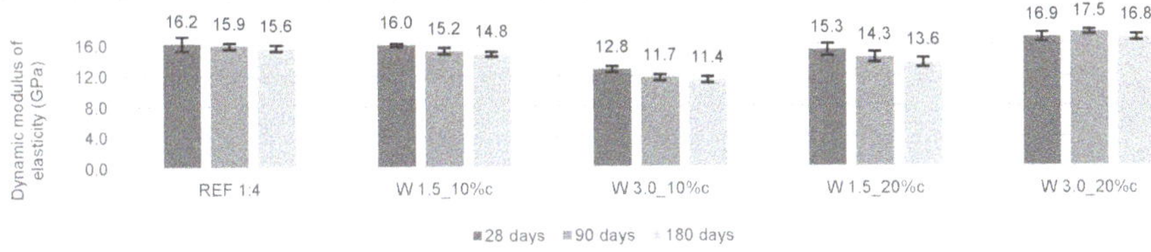

Figure 5. Dynamic modulus of elasticity of the cement mortars.

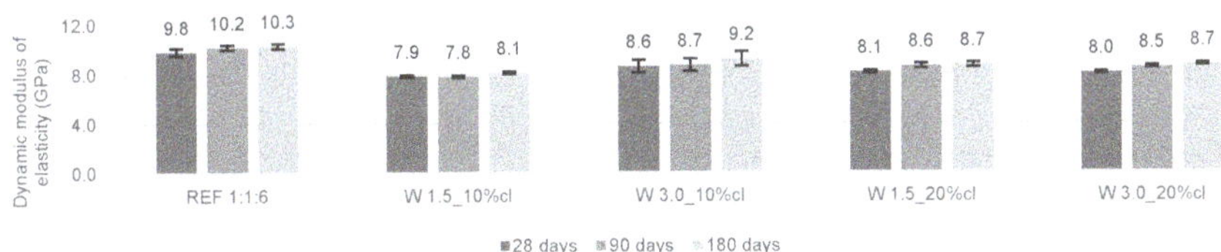

Figure 6. Dynamic modulus of elasticity of the cement-lime mortars.

In general, the modified mortars presented a lower modulus of elasticity when compared to the reference mortars. For cement mortars, the modulus of elasticity decreased from 28 to 180 days. This could be attributed to the internal micro-cracking of the mortars over time. The opposite trend occurred in cement-lime mortars (i.e., the modulus of elasticity presented a slight increase from 28 to 90 days).

According to Araya-Letelier et al. [26], the incorporation of pig hair in mortars did not lead to a significant reduction in the dynamic modulus of elasticity. The authors explained this effect by the small amount of total fiber volume incorporated (up to 1.5%).

3.2.3. Ultra-Sound Pulse Velocity of the Hardened Mortars

The ultra-sound pulse velocity test was performed at 28, 90, and 180 days. The results are presented in Table 5. The ultra-sound pulse velocity results showed that the incorporation of fibers reduced the pulse velocity through the mortar, indicating a decrease in the modulus of elasticity (Table 5).

These results followed the same trend as those of the modulus of elasticity test, as expected. The direct method showed a decrease of the pulse velocity in cement mortars, which could be related to some internal cracks due to shrinkage. According to the indirect method, this reduction is not that significant as it measures the velocity in small distances in the prism, which may detect a more distributed crack pattern and consequent decrease of the pulse velocity, as seen by the high R^2 of the velocity trend lines (Figures 7 and 8).

Table 5. Ultra-sound pulse velocity of the mortars tested.

Mortar	Ultra-Sound Pulse Velocity (m/s)					
	Direct Method			Indirect Method		
	28 days	90 days	180 days	28 days	90 days	180 days
REF 1:4	2855	2751	2661	2551	2581	2676
W 1.5_10%c	2727	2695	2650	2516	2530	2361
W 3.0_10%c	2641	2493	1842	2830	2293	2399
W 1.5_20%c	2556	2602	1855	2677	2739	2545
W 3.0_20%c	2788	2925	1892	2652	2779	2731
REF 1:1:6	2205	2218	2321	2190	2059	2321
W 1.5_10%cl	2050	1990	2051	1921	1968	2003
W 3.0_10%cl	2118	2169	2219	2039	2108	2050
W 1.5_20%cl	2040	2076	2143	2136	2071	2276
W 3.0_20%cl	1985	2065	2120	1957	1936	2017

Figures 5 and 6 present the results of the ultra-sound pulse velocity test by the indirect method for the reference mortars and W 1.5_10% mortars.

3.2.4. Flexural and Compressive Strength of the Hardened Mortars

Figures 9 and 10 present the results of the flexural and compressive strengths at 28 days. In general, the incorporation of wool fibers in cement mortars improved their flexural strength. Longer fibers (3.0 cm) presented a higher increase in flexural strength. The cement mortar with 20% of 3.0 cm long wool fibers had an increase of 40% and 26% in flexural and compressive strength, respectively, compared to REF 1:4.

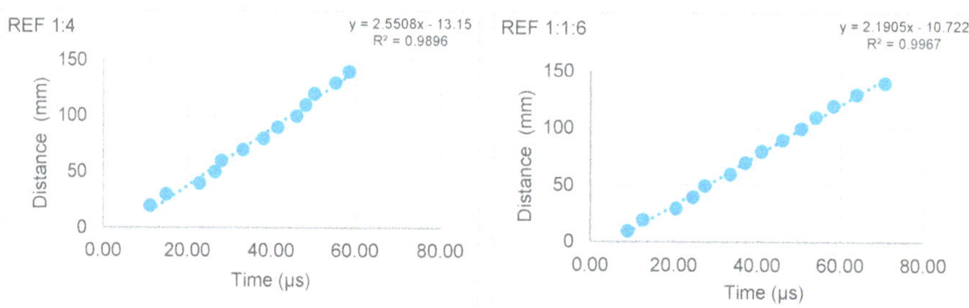

Figure 7. Ultra-sound pulse velocity of the reference mortars.

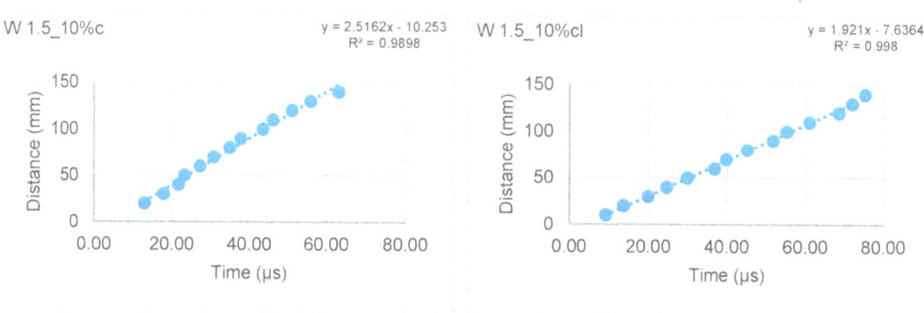

Figure 8. Ultra-sound pulse velocity of the W 1.5_10% mortar.

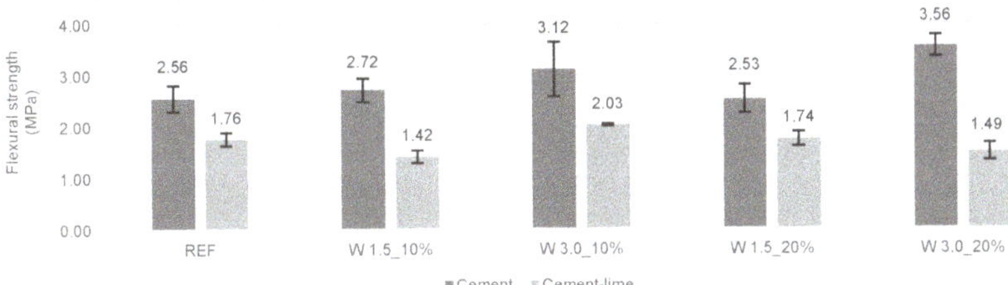

Figure 9. Flexural strength of the mortars at 28 days.

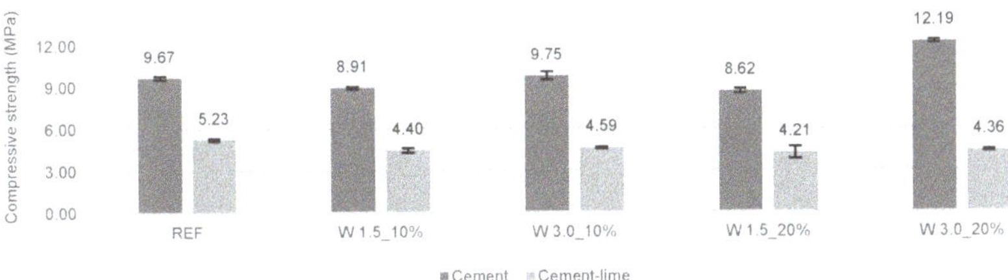

Figure 10. Compressive strength of the mortars at 28 days.

For cement-lime mortars, only W 3.0 cm 10% had an increase of 15% in this property when compared to REF 1:1:6. In general, the modified cement-lime mortars presented a slight decrease in flexural strength. W 1.5 cm 10% obtained the lowest flexural strength, 20% less than the reference mortar. Figure 11 presents the mechanical tests carried out on the mortars.

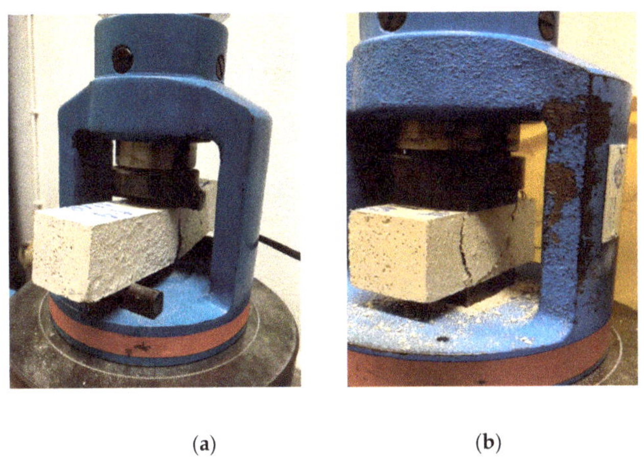

(a)   (b)

Figure 11. Flexural strength test (a); compressive strength test (b).

Fantilli et al. [5] reached similar conclusions to those of the cement mortars. The modified mortars with wool fibers had an 18% higher flexural strength than that of the control mortar. Araya-Letelier et al. [23] studied the incorporation of natural animal fibers such as pig hair in mortars and also reported an increase in flexural strength of the modified mortars. Giosué et al. [11] reported that a hydraulic-lime mortar with wool fibers achieved about a 30% higher flexural strength than that of the conventional mortar. These previous works explained this increase in flexural strength due to a bridging mechanism.

In the compressive strength test, a different trend was found regarding the type of the binder used and the length of the fibers. Indeed, it can be seen that the longer fibers led to an increase in the compressive strength when compared to the cement mortar reference. Modified cement mortars with 1.5 cm of wool fiber presented a slight decrease in compressive strength.

The modified cement-lime mortars obtained a slight reduction of compressive strength compared to the control mortar (REF 1:1:6). W 1.5_20% cl obtained the most significant decrease compared to the reference mortar, about 20%.

Giosué et al. [11] found a decrease in the modified lime-based mortars with wool fibers. The authors reported that the use of fibers reduced the compressive strength of mortars. In agreement with this study, Araya-Letelier et al. [26] found a reduction of compressive strength with the incorporation of animal-based fibers in cement mortars.

Figure 12 presents a sample of the modified mortar, where the wool fibers do not produce a significant change in the mortar's appearance.

Figure 12. Sample of the modified mortar with wool fiber incorporation.

3.2.5. Cracking Behavior

Restrained shrinkage can be induced by the restrictions imposed on deformations of a rendering mortar. Tensile stresses should be dissipated without cracking of the coating. There are several causes that lead the mortar to crack. In order to evaluate the mortar's susceptibility to cracking, some parameters were considered to enhance the analysis of this phenomena as described above, and the results are presented in Table 6.

The values of fracture toughness are presented in Table 6.

An increase in the fracture toughness of the modified cement mortars was noticed. The increment was higher when longer fibers were incorporated, regardless of the incorporation ratio. W 3.0_10%c and W 3.0_20%c attained up to 100% higher toughness values when compared with the reference cement mortar (REF 1:4).

The fracture toughness results of the cement-lime mortars did not present significant changes.

These results are in accordance with previous studies [5,27]. Fantilli et al. [5] also obtained a higher fracture toughness with the incorporation of wool fibers in cementitious composites. This can be explained by the fibers' bridging mechanism, since the fibers cross the micro-cracks, preventing their propagation and delaying the occurrence of the first crack. Reinforced mortars may withstand tensile load after cracking and exhibit ductile behavior [1,28].

Table 6. The mechanical test results of the hardened mortars and parameters related to cracking.

Mortar	Dry Bulk Density (kg/m^3)	Dynamic Modulus of Elasticity (MPa)	Flexural Strength (MPa)	Compressive Strength (MPa)	E/σ_f	σ_f/σ_c	Fracture Toughness (N.mm)
REF 1:4	1919	16210 ± 0.91	2.56 ± 0.21	9.66 ± 0.11	6332	0.27	0.195 ± 0.07
W 1.5_10%c	1906	15980 ± 0.19	2.72 ± 0.19	8.75 ± 0.08	5875	0.31	0.278 ± 0.04
W 3.0_10%c	1845	12790 ± 0.29	3.12 ± 0.43	9.47 ± 0.25	4099	0.33	0.415 ± 0.05
W 1.5_20%c	1879	15290 ± 0.61	2.53 ± 0.22	8.53 ± 0.16	6043	0.30	0.235 ± 0.06
W 3.0_20%c	1899	16860 ± 0.47	3.56 ± 0.17	11.50 ± 0.11	4736	0.31	0.392 ± 0.06
REF 1:1:6	1856	9820 ± 0.21	1.76 ± 0.10	5.22 ± 0.09	5580	0.34	0.135 ± 0.02
W 1.5_10%cl	1808	7850 ± 0.09	1.42 ± 0.09	3.28 ± 0.15	5528	0.43	0.090 ± 0.01
W 3.0_10%cl	1832	8600 ± 0.41	2.03 ± 0.01	4.48 ± 0.06	4236	0.45	0.147 ± 0.03
W 1.5_20%cl	1810	8130 ± 0.05	1.74 ± 0.11	3.17 ± 0.40	4672	0.55	0.127 ± 0.02
W 3.0_20%cl	1807	8030 ± 0.04	1.49 ± 0.13	2.99 ± 0.05	5389	0.50	0.104 ± 0.02

Araya-Letelier et al. [26] found that the incorporation of animal-based fibers, namely pig hair, increased the fracture toughness of the fiber-reinforced mortars. The authors related this increment to the increase in the impact energy absorption capacity of the mortars due to the addition of fibers. In this work, it was reported that the post-cracking behavior of the modified mortars was improved by up to 55% higher energy absorbed at failure.

Considering these factors, all the modified mortars evaluated in this work were less susceptible to cracking. However, the cement-lime mortars presented more ductility than the cement mortars. The incorporation of 20% of 1.5 cm long wool fibers in cement-lime mortars presented the best results regarding ductility. Table 6 presents the results of tests on the hardened mortars.

3.2.6. Open Porosity

The open porosity test determines the volume of interconnected voids in the mortars, in percentage. This property is correlated with the ultra-sound pulse velocity and modulus of elasticity as well as the mechanical strength and water tightness behavior. Table 7 presents the results of the open porosity test.

Table 7. Open porosity test results of the hardened mortars.

Mortar	Open Porosity (%)
REF 1:4	20.18 ± 0.003
W 1.5_10%c	20.91 ± 0.001
W 3.0_10%c	20.91 ± 0.007
W 1.5_20%c	20.82 ± 0.003
W 3.0_20%c	19.88 ± 0.003
REF 1:1:6	23.93 ± 0.003
W 1.5_10%cl	24.93 ± 0.008
W 3.0_10%cl	24.65 ± 0.003
W 1.5_20%cl	24.73 ± 0.008
W 3.0_20%cl	25.49 ± 0.012

The results of the open porosity test confirmed the expectations. In general, the incorporation of fibers increased the volume of pores of the modified mortars. W 3.0_20%c was an exception, as it presented a reduction of 1.5% of total open porosity compared to the control cement mortar. The incorporation of 10% of wool fibers in cement mortars presented the same values, regardless of the length of the fibers.

The cement-lime modified mortars exhibited a greater increase in the volume of pores than that of the cement mortars. W 3.0_20% cl obtained an increase of about 6.5% of total open porosity compared to REF 1:1:6.

Giosué et al. [11] also noticed an increase (32%) of total open porosity in hydraulic lime-mortars with the incorporation of 25% of wool fibers when compared to the control mortar. The increase in open porosity of the modified mortars could be explained by the fiber–matrix interfacial bond that is thought to be less efficient than that of the sand–matrix.

4. Conclusions

From the results of the experimental campaign, it was concluded that the incorporation of wool fibers in rendering mortars presented a satisfactory performance concerning the mechanical and cracking behavior. All of the modified mortars presented less susceptibility to cracking when compared with the mortars without fibers, based on the parameters evaluated.

In general, the reinforced mortars presented a decrease in the dynamic modulus of elasticity, which can be considered an advantage of the incorporation of wool fibers. W 3.0_10%c and W 1.5_10%cl obtained a similar decrease of 20% of modulus of elasticity when compared to the control mortar.

Regarding flexural strength, the modified cement mortars presented an increase when compared to the reference mortar. The addition of longer fibers enhanced the mechanical strength of the mortars. W 3.0_20%c obtained an increase of 40% in flexural strength when compared to REF 1:4. The cement-lime modified mortars obtained a slight reduction in flexural strength, with the exception of W 3.0_10%cl.

The dynamic modulus of elasticity and flexural strength ratio (E/σ_f) was analyzed as an indicator of the mortars' ability to resist cracking. It could be seen that all of the modified mortars, regardless of the binder used, presented a lower (E/σ_f) ratio, which could lead to a lesser tendency to crack due to restrained shrinkage when compared to the reference mortars.

The ratio (σ_f/σ_c) is related to the mortar's ductility. According to the results, all the modified mortars presented a higher ratio (σ_f/σ_c) compared to the reference mortars, which allows concluding that the incorporation of fibers increased the ductility of those mortars, based on this parameter.

Regarding fracture toughness, the modified cement mortars presented improvements. In fact, the toughness increment was higher when longer fibers were incorporated. However, for the cement-lime mortars, the toughness results did not present any significant contribution of the incorporation of the fibers.

Besides the analyzed parameters related with the cracking behavior, it was also found that, concerning compressive strength, the incorporation of longer fibers in cement mortars (W 3.0_10%c and W 3.0_20%c) resulted in an increase.

The results obtained in this work identify the advantages of the addition of the natural wool fibers in rendering mortars, namely concerning the improvement in the mechanical, and in particular of the cracking behavior.

Author Contributions: C.M.P. performed the experiments in the Building Finishes and Thermal Insulation Unit (NRI) of the National Laboratory for Civil Engineering of Portugal (LNEC). The analyses of the tests and the interpretation of the results were developed by C.M.P., R.V. and J.d.B. The original draft of this paper was written by C.M.P. and the review and editing were performed by R.V. and J.d.B.

Funding: This research was funded by Portuguese Foundation for Science and Technology (FCT) (PD/BD/135193/2017).

Acknowledgments: The authors would like to acknowledge the REuSE project from National Laboratory for Civil Engineering of Portugal (LNEC) and the research unit CERIS from Instituto Superior Técnico (IST).

Conflicts of Interest: The authors declare no conflicts of interest.

References

1. Pederneiras, C.M.; Veiga, R.; De Brito, J. Effects of the incorporation of waste fibres on the cracking resistance of mortars: A review. *Int. J. Green Technol.* **2018**, *4*, 38–46.
2. Erdogmus, E. Use of Fiber-Reinforced cements in masonry construction and structural rehabilitation. *Fibers* **2015**, *3*, 41–63. [CrossRef]
3. Grădinaru, C.M.; Bărbuță, M.; Șerbănoiu, A.A.; Babor, D. Investigations of the mechanical properties of concrete with sheep wool fibers and fly ash. *Eng. Sci.* **2016**, *9*, 73–80.
4. Onuaguluchi, O.; Banthia, N. Plant-based natural fibre reinforced cement composites: A review. *Cem. Concr. Compos.* **2016**, *68*, 96–108. [CrossRef]

5. Fantilli, A.P.; Sicardi, S.; Dotti, F. The use of wool as fiber-reinforcement in cement-based mortar. *Constr. Build. Mater.* **2017**, *139*, 562–569. [CrossRef]

6. De Fazio, P.; Cardinale, T.; Arleo, G.; Bernardo, F.; Feo, A. Thermal and mechanical characterization of panels made by cement mortar and sheep's wool fibres. *Energy Procedia* **2017**, *140*, 159–169.

7. Florea, I.; Manea, L. Analysis of thermal insulation building materials based on natural fibers. *Procedia Manuf.* **2019**, *32*, 230–235. [CrossRef]

8. Zach, J.; Korjenic, A.; Petranek, V.; Hroudová, J.; Bednar, T. Performance evaluation and research of alternative thermal insulations based on sheep wool. *Energy Build.* **2012**, *49*, 246–253. [CrossRef]

9. Bosia, D. Sheep wool for sustainable architecture. *Energy Procedia* **2015**, *78*, 315–320. [CrossRef]

10. Alyousef, R.; Aldossari, K.; Ibrahim, O.A.; Mustafa, H.; Jabr, A. Effect of sheep wool fiber on fresh and hardened properties of fiber reinforced concrete. *Int. J. Civ. Eng. Technol.* **2019**, *10*, 190–199.

11. Giosuè, C. Properties of multifunctional lightweight mortars containing zeolite and natural fibers. *J. Sustain. Cem. Mater.* **2019**, *8*, 214–227. [CrossRef]

12. Kesikidou, F.; Stefanidou, M. Natural fiber-reinforced mortars. *J. Build. Eng.* **2019**. [CrossRef]

13. EN 197-1. *Cement Part 1: Composition, Specifications and Conformity Criteria for Common Cements*; European Committee for Standardization (CEN): Brussels, Belgium, 2011.

14. EN 459-1. *Building Lime*; Part 1: Definitions, Specifications and Conformity Criteria; European Committee for Standardization (CEN): Brussels, Belgium, 2015.

15. Cahier 2669-4. *Certification CSTB Des Enduits Monocouches D'imperméabilisation, Modalités D'essais*; Centre Scientifique et Technique du Bâtiment: Marne-la-Vallée, French, 1993.

16. EN 1015-3. *Methods of Test for Mortar for Masonry—Part 3: Determination of Consistence of Fresh Mortar (by Flow Table)*; European Committee for Standardization (CEN): Brussels, Belgium, 1999.

17. EN 1015-6. *Methods of Test for Mortar for Masonry—Part 6: Determination of Bulk Density of Fresh Mortar*; European Committee for Standardization (CEN): Brussels, Belgium, 1998.

18. EN 1015-10. *Methods of Test for Mortar for Masonry—Part 10: Determination of Dry Bulk Density of Hardened Mortar*; European Committee for Standardization (CEN): Brussels, Belgium, 1999.

19. EN 14146. *Natural Stone Test Methods. Determination of the Dynamic Elastic Modulus of Elasticity (by Measuring the Fundamental Resonance Frequency)*; European Committee for Standardization (CEN): Brussels, Belgium, 2004.

20. EN 12504-4. *Testing Concrete in Structures. Part 4: Determination of Ultrasonic Pulse Velocity*; European Committee for Standardization (CEN): Brussels, Belgium, 2007.

21. EN 1015-11. *Methods of Test for Mortar for Masonry—Part 11: Determination of Flexural and Compressive Strength of Hardened Mortar*; European Committee for Standardization (CEN): Brussels, Belgium, 1999.

22. EN 1936. *Natural Stone Test Methods. Determination of Real Density and Apparent Density and Total and Partial Open Porosity*; European Committee for Standardization (CEN): Brussels, Belgium, 2007.

23. Veiga, R. Performance of Rendering Mortars. Contribution to The Study of Their Cracking Resistance. Ph.D. Thesis, University of Porto, Porto, Portugal, 1998. (In Portuguese)

24. Farinha, C.; De Brito, J.; Veiga, R. Incorporation of fine sanitary ware aggregates in coating mortars. *Constr. Build. Mater.* **2015**, *83*, 194–206. [CrossRef]

25. Santos, A.R. The Influence of Natural Aggregates on the Performance of Replacement Mortars for Ancient Buildings: The Effects of Mineralogy, Grading and Shape. Ph.D. Thesis, University of Lisbon, Lisbon, Portugal, 2019.

26. Araya-Letelier, G.; Antico, F.C.; Carrasco, M.; Rojas, P.; García-herrera, C.M. Effectiveness of new natural fibers on damage-mechanical performance of mortar. *Constr. Build. Mater.* **2017**, *152*, 672–682. [CrossRef]

27. Fantilli, A.P.; Jozwiak-niedzwiedzka, D.; Gibas, K.; Dulnik, J. The compatibility between wool fibers and cementitious mortars. In Proceedings of the International Conference on Bio-based Building Materials, Clermont-Ferrand, France, 21–23 June 2017; pp. 42–47.

28. Pereira, M.V.; Fujiyama, R.; Darwish, F.; Alves, G.T. On the strengthening of cement mortar by natural fibers. *Mater. Res.* **2015**, *18*, 177–183. [CrossRef]

Article

Environmentally Sustainable Cement Composites Based on End-of-Life Tyre Rubber and Recycled Waste Porous Glass

Andrea Petrella *, Rosa Di Mundo, Sabino De Gisi, Francesco Todaro, Claudia Labianca and Michele Notarnicola

Dipartimento di Ingegneria Civile, Ambientale, Edile, del Territorio e di Chimica, Politecnico di Bari, Via E. Orabona 4, 70125 Bari, Italy; rosa.dimundo@poliba.it (R.D.M.); sabino.degisi@poliba.it (S.D.G.); francesco.todaro@poliba.it (F.T.); claudia.labianca@poliba.it (C.L.); michele.notarnicola@poliba.it (M.N.)
* Correspondence: andrea.petrella@poliba.it; Tel.: +39(0)805963275; Fax: +39(0)805963635

Received: 11 September 2019; Accepted: 8 October 2019; Published: 10 October 2019

Abstract: In this paper, environmentally sustainable cement mortars were prepared with end-of-life tyre rubber (TR) and recycled waste porous glass (PG) as aggregates in order to obtain lightweight products characterized by renewable and not-pretreated materials specifically for indoor applications. The secondary raw materials were added as partial and/or total replacement of the conventional sand aggregate. The resulting lightweight specimens were characterized by rheological, mechanical, thermal, microstructural and wettability tests. Fine tyre rubber aggregates affected the cohesiveness of the composites, as opposite to coarse tyre rubber and porous glass. The flexural and the compressive strengths of the porous glass samples were higher than the tyre rubber samples because of the higher stiffness and good adhesion of the glass to the cement paste as observed by microstructural observations. On the contrary, an unfavorable adhesion of the tyre aggregates to the cement paste was observed, together with discrete cracks after failure without separation of the two parts of the specimens. The latter result can explain the best results obtained by tyre rubber mortars in the case of impact compression tests where the super-elastic properties of the elastomeric material were evidenced by a deep groove before complete failure. Moreover, the thermal conductivity decrease of the lightweight porous TR and PG composites was in the range of ~80–90% with respect to the sand-based samples, which suggests that they can be used as plasters and masonries, and, in the case of tyre rubber specimens, outside applications are not excluded as observed from the wettability tests.

Keywords: cement composites; recycled waste porous glass; end-of-life tyre rubber; safe production; thermal insulation; mechanical resistance

1. Introduction

The recycling of industrial by-products is an ever rising issue in the sustainable waste management field. Indeed, many studies have focused on the conversion, after appropriate procedures, of these secondary raw materials into a new resource that would otherwise be landfilled. In this respect, over the last years, industrial waste recycling and reuse have become important environmental challenges that many countries are facing in order to reduce overall costs and negative environmental impacts [1–10].

Waste glass and tyre rubber are among the most recycled secondary raw materials from industrial and municipal activities, accordingly, various investigations, mainly in sustainable construction technology, have been carried out with the aim to face the problems relative to the disposal space limitations of these by-products associated with the increasing costs [8,11–17].

In recent years, the amount of waste glass has gradually increased due to the widespread development of urban areas and industries. After the forming process, different types of glass products can be obtained, specifically container glass, flat glass, light bulbs, fluorescent and cathode ray tubes. The biopersistence and chemical inertia of these products lead to long-term accumulation. Accordingly, in the ambit of environmentally sustainable management policies, glass needs to be reused and/or recycled. This material can be indefinitely recycled by preserving the original properties [18,19], and accordingly, the products collected during the sorting operations can be used for the production of abrasives, reflective paints for highways, cullets in glass production, lubricants, additives, fractionators, in road beds and fiberglass production [19–23].

In the last 20 years, recycled glass has been widely used in construction materials, specifically in bricks, normal concrete, pavement materials and asphalt concrete [19,24–28], which is a practice that reduces landfill operations and consumption of natural resources, and also minimizes greenhouse emissions [18,29]. In this respect, it is worth saying that in the case of building materials, recycled glass as aggregate in structural concrete is still not widely used.

The growing amount and disposal of waste tyre rubber has become an environmental issue in many countries. Every year, millions of end-of-life tyres are discarded all over the world and stockpiled tyres represent a threat to human health and the environment through air, water and soil pollution, with its associated economic and social risks [15,30]. The volume of waste, which is globally produced, makes management of the accumulated rubber very hard, with potential fire risks. Tyre burning, although easy and cheap, represents an extremely dangerous method of disposal because fires are difficult to be extinguished and uncontrolled emission of hazardous compounds and potentially toxic gases are released in air. This is very dangerous to humans, animals and plants, and causes ground and surface water contamination generated by the oils and residue ashes left after burning [31,32]. Tyre rubber may be also used as fuel after carbon black production but this solution is not economically advantageous because this material has lower quality and higher costs as compared to conventional fuels [15].

Due to the biopersistence and chemical inertia of waste rubber, recycling operations are rising issues in the sustainable waste management field, as an alternative to landfilling, along with the awareness that new products can be produced with different properties with respect to the original materials.

Tyre rubber can be used for applications in civil and non-civil engineering, for example in erosion control, earthquake shock-wave absorption, road construction as a modifier in asphalt paving mixtures, in breakwaters, in crash and sound barriers, in reefs, playground equipment, as a fuel in cement kilns or for electricity production after incineration [14,16,33–35].

Over the last few years, waste tyre rubber incorporation into cement concrete has been considered one of the most effective, cheap and eco-friendly recycling solutions because it contributes to reducing the cost of some natural aggregates, the great volume of tyre waste, and the emission of toxic compounds and carbon dioxide by preventing tyre fires [36–39].

The main purpose of the present research was to prepare and characterize, by physico-mechanical procedures, eco-friendly non-structural cement composites based on inorganic and organic by-products of recycled waste porous glass (PG) and end-of-life tyre rubber (TR). The secondary raw materials were added as partial and/or total replacement of the conventional sand aggregate, which was made on a volume basis rather than on a weight basis due to the low specific weight of both waste materials. The specimens were characterized by rheological, mechanical, thermal, porosimetric, microstructural and wettability tests. The aim was to obtain lightweight thermo-insulating composites specifically for indoor applications in perfect agreement with the current policies of environmental sustainability. They are also cost-effective because they are prepared through a cheap process where the renewable aggregates are not pre-treated (no addition of chemicals to improve adhesion to the cement paste) and the mixture preparation does not require complex manufacturing processes or expensive procedures.

2. Experimental Part

2.1. Materials and Mortar Specimens Preparation

CEM II A-LL 42.5 R (limestone Portland cement, $R_{c\,(2\,days)} > 25.0$ MPa, $R_{c\,(28\,days)} > 47.0$ MPa, 3100–4400 cm^2/g Blaine specific surface area) was provided by Buzzi Unicem (Barletta, Italy) and used for the preparation of the cement mortars [40]. Conventional sand (normalized) was characterized as clean, isometric and rounded in shape grains in the 0.08–2 mm size range (1660 kg/m^3) and provided by Societè Nouvelle du Littoral, Leucate, France [41,42]. End-of-life tyre rubber (TR) (0–0.5 mm and 0.5–2 mm size range, 460 kg/m^3 and 500 kg/m^3, respectively) and recycled porous waste glass (PG) (0.5–2 mm size range, 300 kg/m^3) were provided by Maltek Industrie S.r.l., Terlizzi, Bari, Italy. PG is a sodium calcium silicate glass (71% SiO_2, 9% CaO, 14% Na_2O, 3% Al_2O_3, 2% MgO, 1% K_2O) obtained from separate collection and separation of municipal and industrial solid wastes. Preliminary cleaning and crushing of the raw materials is followed by the addition of a porosizing agent at temperatures of 900–1300 °C, which induces a controlled porosity of the resulting beads, thus showing a specific weight in the 200–900 kg/m^3 range. TR and PG were added as partial and/or total replacement of the conventional aggregate, which was made on a volume basis rather than on a weight basis due to the low specific weight of both waste materials. In the present case, the total volume of aggregate was set at 500 cm^3 in order to preserve an acceptable workability of the mixture. For this purpose, another sand reference (sand, sample 2) with the same aggregate volume (500 cm^3) and with a 0.5–2 mm sand size range (1880 kg/m^3) was prepared. Tables 1 and 2 report the aggregate types used for the mortar preparation and the composition of the conglomerates. In the present case, the composites were prepared with a water/cement ratio equal to 0.5, specifically with 225 g of water and 450 g of cement; dosages that were chosen according to the standard protocol [41] for the normalized mortar preparation, showing a plastic behavior. After the mixture, the rheology of the fresh mixtures was evaluated by the flow-test [43]. The mortars were placed inside a truncated cone shape ring. After demolding, fifteen hits in fifteen seconds were applied. Flow data were calculated through the following empirical equation after evaluation of the diameters of the mixture before (D_i) and after (D_m) the test:

$$\% flow = \frac{[(D_m - D_i)]}{D_i} * 100 \tag{1}$$

The percentage increase of the diameter of the non-consolidated sample over the base diameter represents the flow of a specimen.

Successively, all the specimens were molded in the form of prisms (40 mm × 40 mm × 160 mm) for the flexural and compressive tests and cured in water for 7, 28, 60 and 90 days after demolding [41]. Moreover, the specimens were molded in the form of cylinders for thermal (φ = 100 mm; H = 50 mm) and impact resistance (φ = 150 mm; H = 60 mm) tests and cured in water for 28 days after demolding.

Table 1. Mortars composition.

Sample	Cement (g)	Water (g)	Sand Volume (cm^3)	TR$_f$ Volume (cm^3)	TR$_c$ Volume (cm^3)	PG Volume (cm^3)
Norm	450	225	810	0	0	0
Sand	450	225	500	0	0	0
TR$_f$	450	225	0	500	0	0
TR$_c$	450	225	0	0	500	0
PG	450	225	0	0	0	500
TR$_f$/Sand	450	225	250	250	0	0
TR$_f$/TR$_c$	450	225	0	250	250	0
TR$_f$/PG	450	225	0	250	0	250

Porosimetric measurements of the resulting mortars were carried-out by Ultrapyc 1200e Automatic Gas Pycnometer, Quantachrome Instruments, Boynton Beach, FL, USA. In this respect, helium gas penetrates the finest pores of the material and the results were the average of three measurements performed on three specimens of the same type (see Table 2).

Table 2. Type, aggregate composition, specific weight ρ and porosity of the cement mortar specimens. Samples prepared with 225 g of water and 450 g of cement. TR_f = fine tyre rubber, TR_c = coarse tyre rubber, PG = porous glass.

Sample	Type	Aggregate Composition	ρ Kg/m^3	Porosity %
1	Norm	Normalized sand	1980	22
2	Sand	100% Sand (0.5–2 mm)	1900	25
3	TR_f	100% TR (0–0.5 mm)	1060	46
4	TR_c	100% TR (0.5–2 mm)	1080	47
5	PG	100% PG (0.5–2 mm)	840	57
6	TR_f/Sand	50% TR (0–0.5 mm)/50% Sand (0.5–2 mm)	1380	40
7	TR_f/TR_c	50% TR (0–0.5 mm)/50% TR (0.5–2 mm)	1080	45
8	TR_f/PG	50% TR (0–0.5 mm)/50% PG (0.5–2 mm)	980	52

2.2. Microscopical Characterization

A scanning electron microscope (SEM) was used to show magnified images of the aggregates and of the cement composites. For this purpose, a FESEM-EDX Carl Zeiss Sigma 300 VP (Carl Zeiss Microscopy GmbH, Jena, Germany) electron microscope was used and the samples were sputtered with gold after immobilization onto aluminum stubs (Sputter Quorum Q150 Quorum Technologies Ltd, East Sussex, UK). The elemental composition of the different organic and inorganic areas of the samples was obtained by energy dispersive X-ray (EDX) analysis (Oxford Instruments, X-Max 20, Abingdon-on-Thames, UK). Specifically, sand composition was: C (4%), O (52%), Si (35%), Ca (2%), end-of-life tyre rubber composition was: C (25%), O (70%), S (1.5%), recycled porous waste glass was: Na (14.8%), Mg (2.1%), Al (3.5%), Si (66%), K (1%), Ca (12%), cement paste composition was: C (4.2%), O (40%), Si (7.6%), Ca (44%), Fe (1.5%), Al (2.5%). A homemade system (premier series dyno-lyte portable microscope and background cold lighting) allowed us to evaluate the wettability of the specimens, which was carried out after deposition of a drop of water onto the side and fracture surface of each sample.

2.3. Mechanical and Thermal Tests

The flexural and compression tests were carried-out by the use of a MATEST device, Milan, Italy. Compression strengths were obtained on twelve semi-prisms (loading rate in the range of 2400 ± 200 N/s), deriving from the flexural tests carried-out on six prisms (40 mm × 40 mm × 160 mm) (loading rate in the range of 50 ± 10 N/s) [41].

An impact resistance device was made according to the ACI Committee 544 [44]. Specifically, the test was carried-out by dropping a weight of 4.50 kg from a height of 45 cm above a steel ball (63 mm diameter) placed centrally on the upper surface of the specimen. The energy absorbed by the sample before the fracture was obtained after evaluation of the number of blows on the sphere.

Thermal conductivity (λ) and thermal diffusivity (α) measurements were carried-out through an ISOMET 2104 device, from Applied Precision Ltd. (Bratislava, Slovakia). Before the tests, the mortars were dried to a constant weight, followed by final stabilization at room temperature. Specifically, a constant thermal flow was produced by a heating probe that was applied on the sample surface and the temperature was recorded over time. The thermal diffusivity and thermal conductivity parameters were obtained by comparison between the experimental temperature values and the analytical solution of the heat conduction equation [45].

3. Results and Discussion

3.1. Characterization of the Aggregates and Rheological Tests of the Mortars

Figure 1 shows the scanning electron micrographs (SEM) of tyre rubber grains and of PG beads. An intrinsic micro-scale texture of the elastomeric aggregates can be observed (Figure 1A), while the glass aggregates (Figure 1B) show a large open porosity together with a large closed porosity (inset Figure 1B). The properties of these secondary raw materials can explain the properties of the resulting mortars. In this respect, Table 2 shows that the TR and PG samples are lighter and with a much higher porosity than the references, while, among the lightweight composites, PG mortars were the most porous.

Figure 1. (**A**) Tyre rubber (TR) grain and (**B**) porous glass (PG) bead with evidenced porosity (in the inset: inner porosity).

Flow-test measurements were carried-out on the mixtures in order to determine the consistency of the fresh specimens (Figure 2).

Sample 2 (sand sample) showed a higher flow (+56%) than the control (normalized mortar (norm), with a plastic behavior) due to the lack of fines in the sand aggregate. The sample with bare fine tyre rubber aggregates (sample 3) showed a flow decrease in the range of 44.5% with respect to the reference [46], as opposite to sample 4 with bare coarse tyre rubber aggregates, which showed approximately the same flow as the control (+9%). The former result is ascribed to the absence of the fine aggregates, with a higher specific surface, which contributes to the decrease of plasticity and increase of cohesiveness of the specimen. Sample 6, with fine TR and coarse sand, showed a plastic behavior as the flow was similar to the control (+8%), while the PG composite (sample 5) showed a flow increase in the range of 30% because of the absence of fine aggregates. Finally, the presence of the finer TR fraction is associated with the decrease of workability of the samples 7 and 8 (−25%).

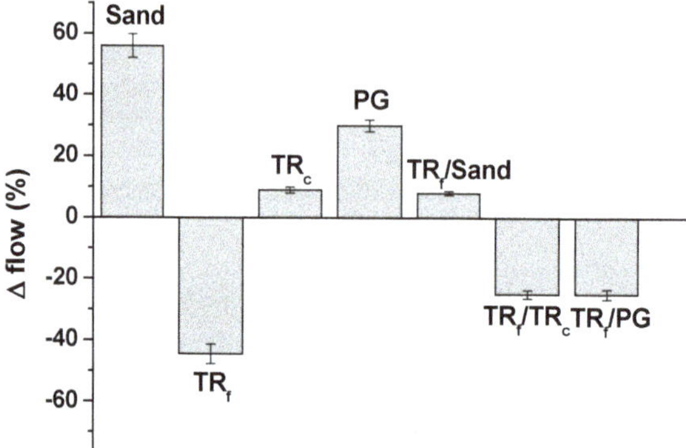

Figure 2. Flow-test results with respect to the normalized mortar (norm sample).

3.2. Mechanical Tests and Microscopical Characterization of the Mortars

Figures 3 and 4 report the flexural and compressive strengths of the samples. Reference sand-based mortars (samples 1 and 2) showed the best mechanical performances. In fact, the flexural and compressive strengths of all the unconventional and lightweight mortars were lower than the references, based on the more resistant sand aggregate.

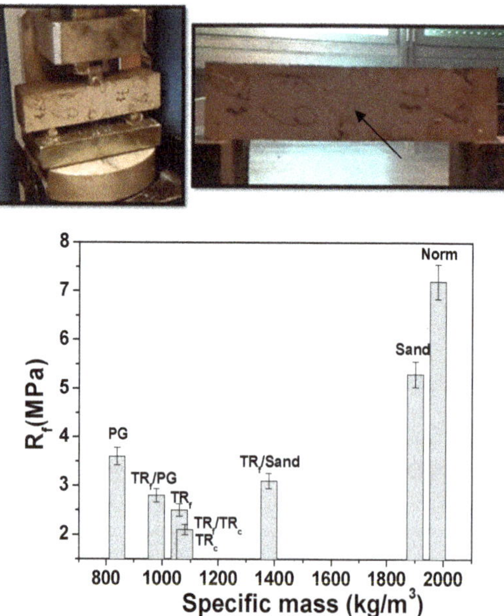

Figure 3. Flexural strengths of the cement mortars. On the top: discrete cracks after rupture in the TR specimens (evidenced by the arrow), with the two parts of the sample still connected by the tyre rubber.

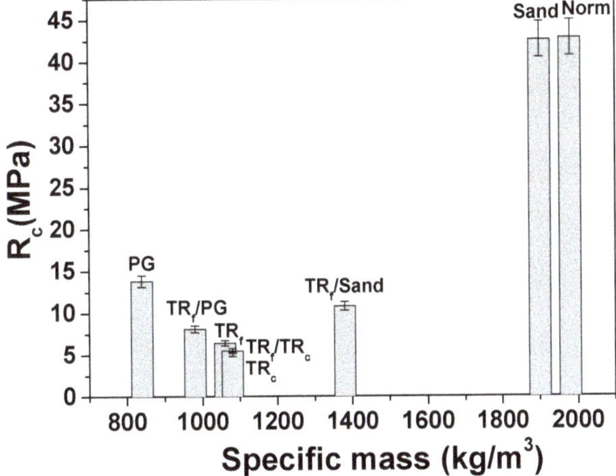

Figure 4. Compressive strengths of the cement mortars.

As for specimens 3, 4 and 7, characterized by bare tyre rubber, the addition of the elastomeric waste disrupted the mineral skeleton of the mortars after the formation of voids in the composite, which sensibly reduced the specific weight of the samples.

This effect depends on the organic/inorganic interface properties. In fact, an unfavorable adhesion of the aggregate to the cement paste [46–52] was observed and ascribed to the hydrophobic nature of the tyres rubber and to the completely different chemistry of the compounds present in the polymeric structure and in the inorganic matrix (Figure 5A). This evidence was not observed in the sand-based references where a good aggregate/cement paste adhesion was present (Figure 5B).

Accordingly, after total replacement of the sand volume, a decrease in the mechanical performances of these composites was ascribed to the low density of the TR grains and to the voids (entrapped air) created by the aggregate at the cement/TR interface during mixing, which can explain the decrease of the specific weight and an almost double porosity of these samples with respect to the references [46–52] (Table 2).

Specifically, the flexural resistances of samples 3, 4 and 7 were 60–70% lower than the references, while the compressive resistances were ~85% lower than the references (Figures 3 and 4). Replacement of 50% of the sand volume with TR grains (sample 6) led to an increase in the mechanical resistances with respect to the composites with 100% sand replacement, due to the presence of the more resistant sand aggregate. In fact, in this case, the flexural strength decrease was approximately 40–55% with respect to both references, while the compressive resistances were ~70% lower than the reference samples [51,53–57]. The specimen with fine tyre rubber aggregates (samples 3) showed higher mechanical resistances than the coarse tyre rubber type (samples 4), as was also observed in previous works [53–56], a result ascribed to the higher surface area of the fine type elastomeric materials, which, as observed in rheological measurements, improve the cohesiveness of the mixture. Mortars based on PG and PG/TR aggregates (samples 5 and 8) showed an increase in the flexural and compressive strengths with respect to tyre rubber composites, in particular the flexural resistance of sample 5 (PG specimen) was almost double the other tyre rubber samples (samples 3, 4 and 7) together with a three times higher increase of the compressive strength. Moreover, the PG composites showed a decrease in the flexural resistances in the range of 25–50% (sample 5) and 50–60% (sample 8) with respect to the references, and a decrease in the compressive resistances in the range of ~65% (sample 5) and ~80% (sample 8) with respect to the references [58–60]. The presence of tyre rubber was detrimental for the mechanical strengths, which conversely were interesting when bare PG was used because glass showed higher stiffness and better adhesion to the cement paste due to the high roughness of the beads and to a chemical composition (silicates, aluminates) similar to the ligand matrix (Figure 5C) [11,61]. For this reason, the lowest specific weight of the glass samples was exclusively ascribed to the intrinsic porosity of the aggregate and not to the porosity of the composite at the interface. Table 3 shows the results obtained from the mechanical tests.

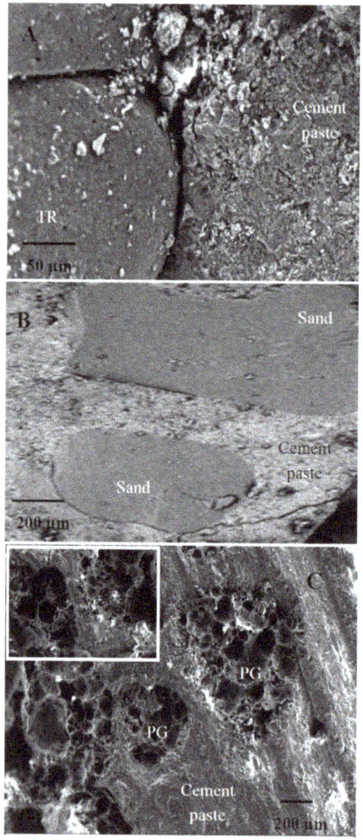

Figure 5. SEM images of: (**A**) cement/TR interface, (**B**) cement/sand interface, (**C**) cement/PG interface.

The flexural failure mode of the mortars containing bare TR aggregate did not exhibit the typical brittle behavior observed in the conventional sand-based samples (samples 1 and 2), indeed a separation of the two parts of the specimens was not observed but only discrete cracks, ascribed to the tyre tensile strength (Figure 3, on top) [53,62].

Similar to flexural strength observations, the compressive failure observed in the case of TR mortars was more gradual and the specimens showed a high-energy absorption capacity because of the load retention after failure without collapse. It was also observed that many aggregates of the TR specimens sheared off along the failure plane, and accordingly, the bond between the TR aggregate and the cement paste was stronger than the failure strength of the aggregate granules. Reference samples and sample 5 (based on bare PG) showed a typical brittle failure [60,63–65]. A semi-brittle failure was observed in the case of the Sand-TR and of the PG/TR samples (sample 6 and 8). The R_c values of the PG, TR_f/Sand and TR_f/PG samples were in the CS IV conformity range for plasters, while the R_c values of the TR_f, TR_c and TR_f/TR_c samples were in the CS III conformity range for plasters [66]. The R_c values of the PG and TR_f/Sand samples were in the M10 conformity range for masonries, while the R_c values of TR_f/PG, TR_f, TR_c and TR_f/TR_c were in the M5 conformity range for masonries [67].

Table 3. Main results from flexural (R_f), compressive (R_c), impact resistance (IR) and thermal tests (λ).

Sample	$R_{f\,(28d)}$ (MPa)	$R_{c\,(28d)}$ (MPa)	IR (J/cm^2)	λ (W/mK)
Norm	7.2	42.8	4.5	1.95
Sand	5.3	42.6	3.2	1.6
TR$_f$	2.5	6.4	51.6	0.3
TR$_c$	2.1	5.1	19.1	0.3
PG	3.6	13.8	1.3	0.2
TR$_f$/Sand	3.1	10.8	6.4	0.7
TR$_f$/TR$_c$	2.1	5.5	22.3	0.25
TR$_f$/PG	2.8	8.1	11	0.3

Figure 6 shows a picture of the sections of the specimens after the mechanical tests, where the nature of the aggregates can be observed together with a good distribution of the tyre rubber grains, porous glass beads and conventional sand. The samples, although characterized by different types of aggregates and with different grain size distribution, were extremely homogenous without any form of segregation.

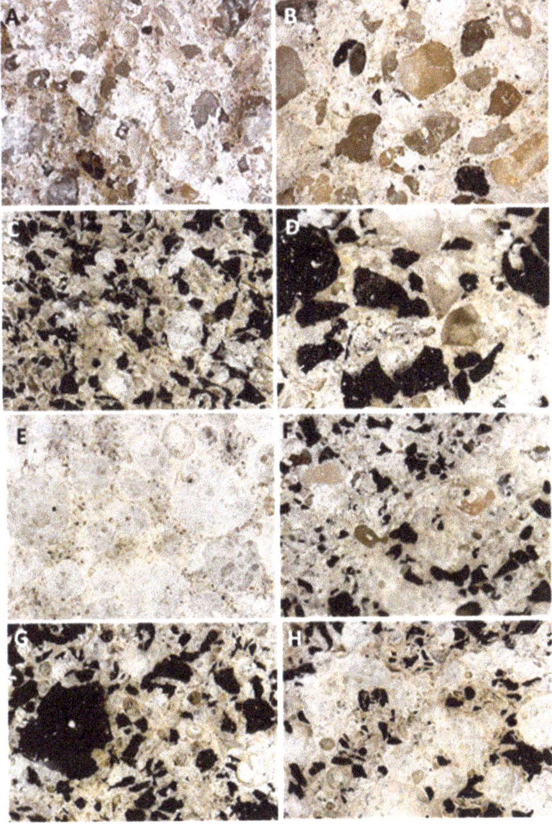

Figure 6. Sections of the specimens after the mechanical tests. (**A**) Norm (sample 1), (**B**) Sand (sample 2), (**C**) TR$_f$ (sample 3), (**D**) TR$_c$ (sample 4), (**E**) PG (sample 5), (**F**) TR$_f$/Sand (sample 6), (**G**) TR$_f$/TR$_c$ (sample 7), (**H**) TR$_f$/PG (sample 8).

The temporal evolution of the flexural and compressive resistances of samples 3, 4, 5 and 8 is reported in Figure 7. An increase in the strengths can be observed for every composite with a final stabilization in the range of 60–90 days. This result may demonstrate a stability of the materials, in consideration of the adopted curing/conservation conditions of the conglomerates (in water).

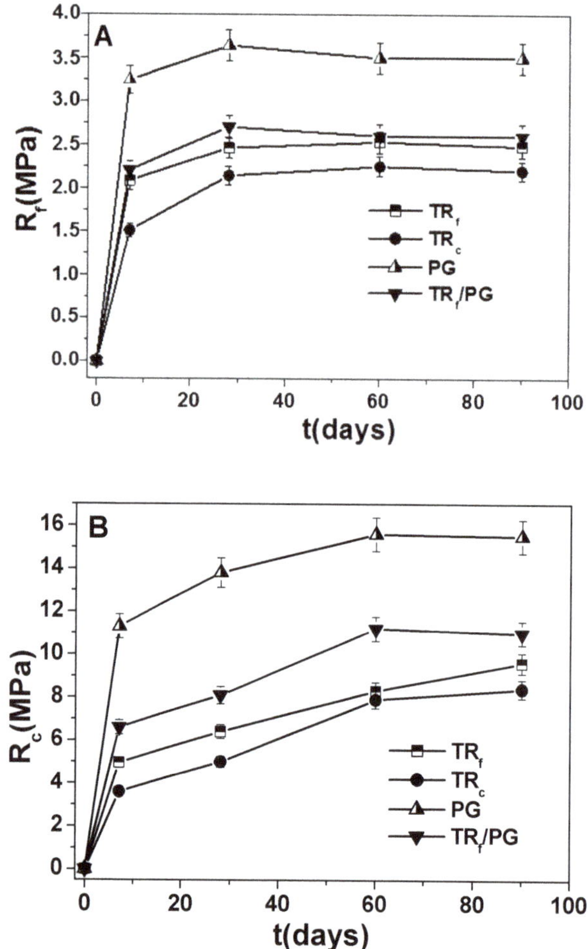

Figure 7. Temporal evolution of the (**A**) flexural and (**B**) compressive resistances of sample 3 (TR_f), sample 4 (TR_c), sample 5 (PG) and sample 8 (TR_f/PG).

From the impact compression tests (Figure 8A), it was observed that the references and the PG mortars (samples 1, 2 and 5) were extremely fragile, with a relatively low number of blows necessary for fracture formation (Figure 8D,E).

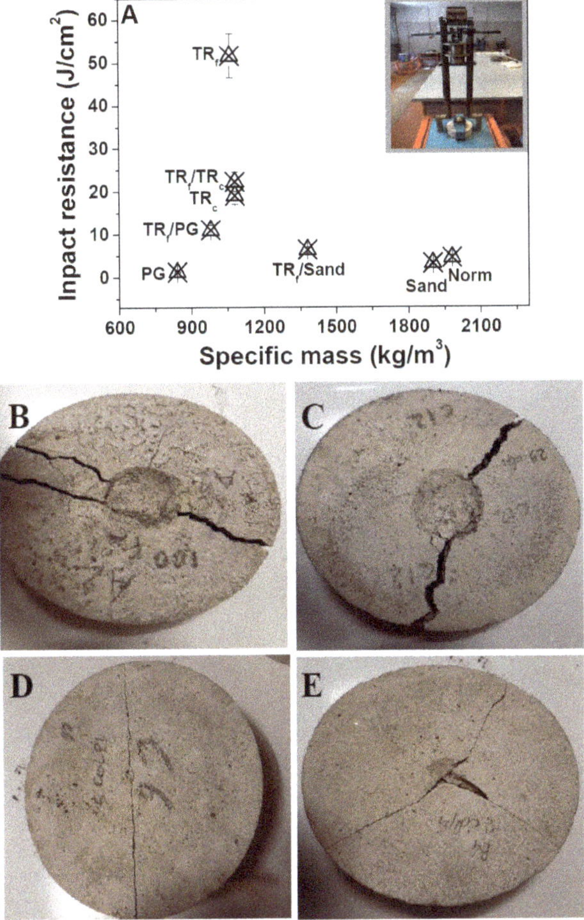

Figure 8. (**A**) Impact resistance of the cement mortars, (**B**) TR$_f$ (sample 3), (**C**) TR$_f$/TR$_c$ (sample 7), (**D**) Norm (sample 1), (**E**) PG (sample 5).

The best results were obtained with the TR specimens because of the load retention of these composites, ascribed to the super-elastic properties of the elastomeric material and evidenced by a deep groove before complete failure (Figure 8B,C) [46,68]. Specifically, sample 3 with finer grains showed the highest energy absorption capacity ascribed to a better compaction with respect to the other similar composites. As for the flexural and compressive tests, average values were observed in samples with 50% of TR (samples 6 and 8) because of the presence of brittle materials as sand and PG.

3.3. Thermal Tests

TR-based mortars (samples 3, 4 and 7) showed lower thermal conductivities and diffusivities (80–85%) as compared to the sand equivalent controls (Figure 9) because of the lower specific weight of the specimens due to the polymer characteristics (low specific weight of the aggregate) and also to the voids (entrapped air) at the TR/cement paste interface, which limit heat transport through the material (see Figure 5A) [50,68]. The best results were obtained in the case of the PG mortar (0.2 W/mK, sample 5). Specifically, a corresponding decrease (~90%) of the thermal conductivity with respect to the controls was observed, a result ascribed to the large porosity of the glass beads, which induces a further increase in the thermal insulation (see Figure 1B). In addition, the TR/PG mixture affected the thermal insulation. Average values (60–65%) were obtained in samples with the presence of 50% of sand (sample 6). An exponential decrease in the conductivity and diffusivity data was observed with the decrease in the conglomerates specific weight. Table 3 also shows the results obtained from the thermal tests.

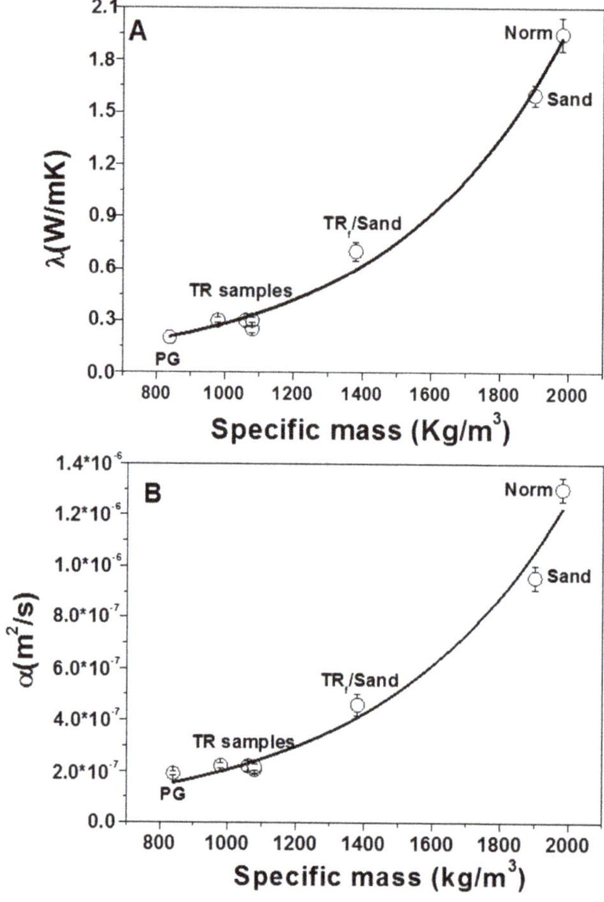

Figure 9. (**A**) Thermal conductivity and (**B**) thermal diffusivity of the cement mortar specimens.

3.4. Wettability Tests

An investigation on the wettability of the tyre rubber specimens was carried out in a previous work [52]. In the present paper, it was completed with the comparison of the properties of the porous glass mortars. As known, the wettability is the ability of a liquid to maintain contact with a solid surface, thus a hydrophobic behavior is associated with surfaces that repel water (poor wettability), while a hydrophilic behavior is associated with a favorable wettability of the surfaces [69]. A surface is considered hydrophobic if the water contact angle is higher than 90°, whereas a surface is considered hydrophilic if the water contact angle is lower than 90°. As has also been formerly observed [52], the surface and the bulk of the sand-based samples (norm and sand) showed an average fast water absorption and a hydrophilic behavior (water contact angle lower than 90°) due to the hydrophilic porous domains of the cement paste (Figure 10A). Similar results were obtained in the case of the porous glass mortar (PG), a result ascribed to the hydrophilic nature and the high porosity of the soda-lime aggregate (Figure 5C), together with the presence of the hydrophilic porous domains of the cement paste, which determine a fast penetration of water (Figure 10B). Tyre-rubber specimens (TR$_f$, TR$_c$, TR$_f$/TR$_c$) showed a strong reduction in water penetration both on the surface and on the bulk and a hydrophobic behavior (water contact angle higher than 90° [52]), although these samples were more porous than the references (poor adhesion of the aggregate to the cement paste, see Figure 5A). These results were totally ascribed to the hydrophobic nature of the organic aggregate. Maximum hydrophobic performances were obtained in the presence of the finer tyre rubber grain size distribution (TR$_f$, Figure 10C). In addition, in the case of the rubber/sand sample (TR$_f$/Sand) the water absorption was significantly lower than the reference samples (~15% on the side surface, ~25% on the fracture surface), but higher than the TR-mortars due to the halved volume of rubber, which dramatically reduced the net force for water penetration and thus stabilized the deposited drops on the surface [52]. Interestingly, the rubber/glass sample (TR$_f$/PG) showed a low water absorption (~10% on the side surface, ~15% on the fracture surface) ascribed to the contribution of the organic hydrophobic aggregate in spite of the contribution of the hydrophilic and porous glass (Figure 10D). Therefore, in this case, the rubber contributed to the reduction of the net force for water penetration, thus stabilizing the deposited drops on the surface.

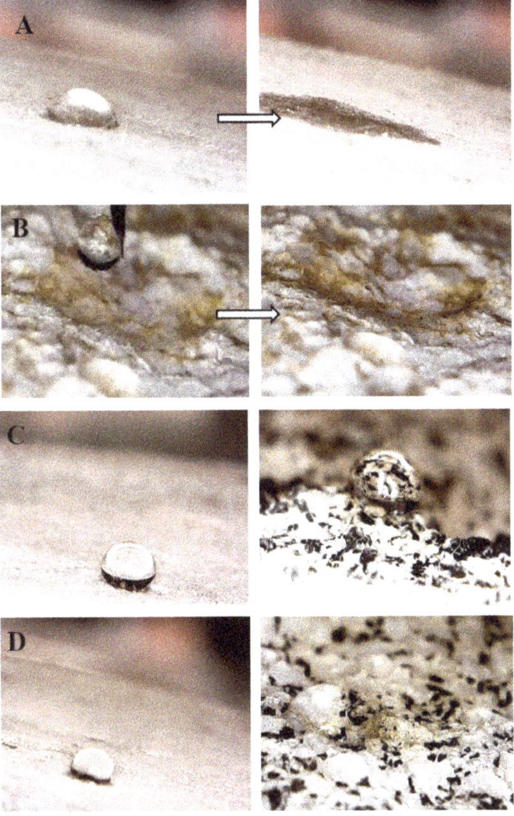

Figure 10. Wettability tests on the side surface of (**A**) Norm (sample 1) at t = 0 s (left) and at t = 60 s (right), on the fracture surface of (**B**) PG (sample 5) at t = 0 s (left) and at t = 2 s (right), on the side (left) and fracture (right) surface of (**C**) TR$_f$ (sample 3) at t = 150 s, on the side (left) and fracture (right) surface of (**D**) TR$_f$/PG (sample 8) at t = 150 s.

It is worth saying that all materials showed a similar behavior because the surface and bulk of all the specimens were almost the same, which means that the mortar features cannot be modified by eventual wear or damage events of the surface.

4. Conclusions

End-of-life tyre rubber (TR) and recycled waste porous glass (PG) were employed for the production of lightweight eco-sustainable cement conglomerates specifically for indoor applications. A cheap and environmentally friendly process was used because the aggregates were not pre-treated. The secondary raw materials were added as partial and/or total replacement of the conventional sand aggregate, which was made on a volume basis rather than on a weight basis due to the low specific weight of both waste materials. The specimens were characterized by rheological, mechanical, thermal, microstructural and wettability tests.

TR and PG were added as partial and/or total replacement of the conventional aggregate, which was made on a volume basis rather than on a weight basis due to the low specific weight of both waste materials. In the present case, the total volume of aggregate was set at 500 cm^3 in order to preserve an acceptable workability of the mixture. The samples were prepared with a water/cement ratio equal to 0.5, a value that was chosen according to the standard protocol for the normalized mortar preparation, showing a plastic behavior.

The main results showed that:

(a) Fine TR aggregates affected the cohesiveness of the mixtures as opposite to coarse TR and PG types.

(b) The flexural and compressive strengths of the unconventional and lightweight mortars were lower than the references, based on the more resistant sand aggregate. Specifically, a decrease of ~60% and ~85% respectively of the flexural and compressive strengths was observed in the case of TR samples, whereas a lower decrease was observed in the case of the PG specimens (~25% and ~65% decrease respectively of the flexural and compressive strengths). The flexural and compressive strengths of the PG samples were higher than the TR samples because of the higher stiffness and good adhesion of the glass to the cement paste, ascribed to the roughness of the aggregate surface together with a chemical composition (silicates, aluminates) similar to the ligand matrix. On the contrary, an unfavorable adhesion of the rubber aggregate to the cement paste was observed and ascribed to the hydrophobic nature of the organic material and to the completely different chemical composition of the polymer and of the inorganic matrix.

(c) The specimen with fine tyre rubber aggregates showed higher mechanical resistances than the coarse tyre rubber type, a result ascribed to the higher surface area of the fine type elastomeric materials, which, as observed in rheological measurements, improves the cohesiveness of the mixture. Replacement of 50% of the sand volume with TR grains led to an increase in the mechanical resistances with respect to the composites with 100% sand replacement, due to the presence of the more resistant sand aggregate.

(d) The flexural failure mode of the mortars containing bare TR aggregate did not exhibit the typical brittle behavior observed in the conventional sand-based samples and in the bare PG samples, indeed a separation of the two parts of the specimens was not observed but only discrete cracks were noticed and ascribed to the rubber tensile strength.

(e) From the impact compression tests it was observed that the references and the PG mortars were extremely fragile, and the best results were obtained with the TR specimens because of the load retention of these composites ascribed to the super-elastic properties of the elastomeric material and evidenced by a deep groove before complete failure.

(f) From thermal measurements it was observed that the thermal conductivity and diffusivity decrease of the lightweight materials (tyre rubber and porous glass specimens) was in the range of ~80–90% with respect to the sand-based samples. This was ascribed to the large porosity of the glass beads and, in the case of the TR specimens, to the voids at the TR/cement paste interface, which limit heat transport through the material.

(g) Suitable applications in the construction industry as non-structural artifacts may be found for all the samples, as acceptable compressive data for plasters and masonries were obtained. Specifically, after wettability investigations, bare PG specimens may be suitable for indoor applications, while TR specimens may be also suitable for outside elements exposed for example to water flowing and capillary rise. The latter application may also be indicated for the glass/tyre rubber mortars characterized by hydrophobic behavior and low water absorption, ascribed to the presence of the organic aggregate, and by interesting mechanical resistances and high thermo-insulation, mainly ascribed to the inorganic aggregate.

Finally, it is worth considering that these waste/cement composites are cost-effective and environmentally sustainable construction materials because they are prepared through a cheap and eco-friendly process where the aggregates were not pre-treated and the mixture preparation did not require complex manufacturing processes or expensive procedures.

Author Contributions: Conceptualization, M.N.; methodology, R.D.M.; software, F.T.; validation, C.L.; formal analysis, A.P.; investigation, A.P.; data curation, S.D.G.; writing—original draft preparation, A.P.; writing—review and editing, A.P. and M.N.; supervision, M.N.

Funding: This research received no external funding

Acknowledgments: The authors wish to thank Pietro Stefanizzi and Stefania Liuzzi for thermal analysis. The authors also thank Adriano Boghetich for SEM-EDX analysis. Regione Puglia is gratefully acknowledged for financial support (Micro X-Ray Lab Project–Reti di Laboratori Pubblici di Ricerca, cod. n. 45 and 56). The Department of Civil, Environmental, Land, Construction and Chemistry (DICATECh) of the Polytechnic University of Bari is gratefully acknowledged for SEM analyses.

Conflicts of Interest: The authors declare no conflict of interest.

References

1. Geraldo, R.H.; Pinheiro, S.M.; Silva, J.S.; Andrade, H.M.; Dweck, J.; Gonçalves, J.P.; Camarini, G. Gypsum plaster waste recycling: A potential environmental and industrial solution. *J. Clean. Prod.* **2017**, *164*, 288–300. [CrossRef]

2. Gu, F.; Guo, J.; Zhang, W.; Summers, P.A.; Hall, P. From waste plastics to industrial raw materials: A life cycle assessment of mechanical plastic recycling practice based on a real-world case study. *Sci. Total Environ.* **2017**, *601*, 1192–1207. [CrossRef] [PubMed]

3. Singh, N.; Hui, D.; Singh, R.; Ahuja, I.P.S.; Feo, L.; Fraternali, F. Recycling of plastic solid waste: A state of art review and future applications. *Compos. Part B Eng.* **2017**, *115*, 409–422. [CrossRef]

4. Sienkiewicz, M.; Janik, H.; Borzędowska-Labuda, K.; Kucińska-Lipka, J. Environmentally friendly polymer-rubber composites obtained from waste tyres: A review. *J. Clean. Prod.* **2017**, *147*, 560–571. [CrossRef]

5. Liuzzi, S.; Rubino, C.; Stefanizzi, P.; Petrella, A.; Boghetich, A.; Casavola, C.; Pappalettera, G. Hygrothermal properties of clayey plasters with olive fibers. *Constr. Build. Mater.* **2018**, *158*, 24–32. [CrossRef]

6. Petrella, A.; Spasiano, D.; Acquafredda, P.; De Vietro, N.; Ranieri, E.; Cosma, P.; Rizzi, V.; Petruzzelli, V.; Petruzzelli, D. Heavy metals retention (Pb (II), Cd (II), Ni (II)) from single and multimetal solutions by natural biosorbents from the olive oil milling operations. *Process Saf. Environ.* **2018**, *114*, 79–90. [CrossRef]

7. Leng, Z.; Padhan, R.K.; Sreeram, A. Production of a sustainable paving material through chemical recycling of waste PET into crumb rubber modified asphalt. *J. Clean. Prod.* **2018**, *180*, 682–688. [CrossRef]

8. Yao, Z.; Ling, T.C.; Sarker, P.K.; Su, W.; Liu, J.; Wu, W.; Tang, J. Recycling difficult-to-treat e-waste cathode-ray-tube glass as construction and building materials: A critical review. *Renew. Sustain. Energy Rev.* **2018**, *81*, 595–604. [CrossRef]

9. Coppola, L.; Bellezze, T.; Belli, A.; Bignozzi, M.C.; Bolzoni, F.; Brenna, A.; Cabrini, M.; Candamano, S.; Cappai, M.; Caputo, D.; et al. Binders alternative to Portland cement and waste management for sustainable construction-part 1. *J. Appl. Biomater. Funct. Mater.* **2018**, *16*, 186–202.

10. Coppola, L.; Bellezze, T.; Belli, A.; Bignozzi, M.C.; Bolzoni, F.; Brenna, A.; Cabrini, M.; Candamano, S.; Cappai, M.; Caputo, D.; et al. Binders alternative to Portland cement and waste management for sustainable construction-part 2. *J. Appl. Biomater. Funct. Mater.* **2018**, *16*, 207–221.

11. Petrella, A.; Spasiano, D.; Race, M.; Rizzi, V.; Cosma, P.; Liuzzi, S.; De Vietro, N. Porous waste glass for lead removal in packed bed columns and reuse in cement conglomerates. *Materials* **2019**, *12*, 94. [CrossRef] [PubMed]

12. Kim, K.; Kim, K. Valuable Recycling of waste glass generated from the liquid crystal display panel industry. *J. Clean. Prod.* **2018**, *174*, 191–198. [CrossRef]

13. Paul, S.C.; Šavija, B.; Babafemi, A.J. A comprehensive review on mechanical and durability properties of cement-based materials containing waste recycled glass. *J. Clean. Prod.* **2018**, *198*, 891–906. [CrossRef]

14. Kroll, L.; Hoyer, S.; Klaerner, M. Production technology of cores for hybrid laminates containing rubber powder from scrap tyres. *Procedia Manuf.* **2018**, *21*, 591–598. [CrossRef]

15. Thomas, B.S.; Gupta, R.C. A comprehensive review on the applications of waste tire rubber in cement concrete. *Renew. Sustain. Energy Rev.* **2016**, *54*, 1323–1333. [CrossRef]

16. Presti, D.L.; Izquierdo, M.A.; del Barco Carrión, A.J. Towards storage-stable high-content recycled tyre rubber modified bitumen. *Environments* **2018**, *172*, 106–111. [CrossRef]

17. Ramirez-Canon, A.; Muñoz-Camelo, Y.; Singh, P. Decomposition of used tyre rubber by pyrolysis: Enhancement of the physical properties of the liquid fraction using a hydrogen stream. *Environments* **2018**, *5*, 72. [CrossRef]

18. Sobolev, K.; Turker, P.; Soboleva, S.; Iscioglu, G. Utilization of waste glass in ECO cement, strength properties and microstructural observations. *Waste Manag.* **2006**, *27*, 971–976. [CrossRef] [PubMed]

19. Silva, R.V.; De Brito, J.; Lye, C.Q.; Dhir, R.K. The role of glass waste in the production of ceramic-based products and other applications: A review. *J. Clean. Prod.* **2017**, *167*, 346–364. [CrossRef]

20. Chen, G.; Lee, H.; Young, K.L.; Yue, P.L.; Wong, A.; Tao, T.; Choi, K.K. Glass recycling in cement production—An innovative approach. *Waste Manag.* **2002**, *22*, 747–753. [CrossRef]

21. Sun, Z.; Shen, Z.; Ma, S.; Zhang, X. Sound absorption application of fiberglass recycled from waste printed circuit boards. *Mater. Struct.* **2015**, *48*, 387–392. [CrossRef]

22. Lu, J.X.; Duan, Z.H.; Poon, C.S. Combined use of waste glass powder and cullet in architectural mortar. *Cem. Concr. Comp.* **2017**, *82*, 34–44. [CrossRef]

23. Spasiano, D.; Luongo, V.; Petrella, A.; Alfè, M.; Pirozzi, F.; Fratino, U.; Piccinni, A.F. Preliminary study on the adoption of dark fermentation as pretreatment for a sustainable hydrothermal denaturation of cement-asbestos composites. *J. Clean. Prod.* **2017**, *166*, 172–180. [CrossRef]

24. Lachance-Tremblay, É.; Perraton, D.; Vaillancourt, M.; Di Benedetto, H. Effect of hydrated lime on linear viscoelastic properties of asphalt mixtures with glass aggregates subjected to freeze-thaw cycles. *Const. Build. Mater.* **2018**, *184*, 58–67. [CrossRef]

25. Amlashi, S.M.H.; Vaillancourt, M.; Carter, A.; Bilodeau, J.P. Resilient modulus of pavement unbound granular materials containing recycled glass aggregate. *Mater. Struct.* **2018**, *51*, 89. [CrossRef]

26. Petrella, A.; Petrella, M.; Boghetich, G.; Basile, T.; Petruzzelli, V.; Petruzzelli, D. Heavy metals retention on recycled waste glass from solid wastes sorting operations: A comparative study among different metal species. *Ind. Eng. Chem. Res.* **2012**, *51*, 119–125. [CrossRef]

27. Petrella, A.; Petruzzelli, V.; Ranieri, E.; Catalucci, V.; Petruzzelli, D. Sorption of Pb(II), Cd(II) and Ni(II) from single- and multimetal solutions by recycled waste porous glass. *Chem. Eng. Commun.* **2016**, *203*, 940–947. [CrossRef]

28. Petrella, A.; Petruzzelli, V.; Basile, T.; Petrella, M.; Boghetich, G.; Petruzzelli, D. Recycled porous glass from municipal/industrial solid wastes sorting operations as a lead ion sorbent from wastewaters. *React. Funct. Polym.* **2010**, *70*, 203–209. [CrossRef]

29. Rakshvir, M.; Barai, S.V. Studies on recycled aggregates-based concrete. *Waste Manag. Res.* **2006**, *24*, 225–233. [CrossRef]

30. Zhang, Y.; Hwang, J.Y.; Peng, Z.; Andriese, M.; Li, B.; Huang, X.; Wang, X. Microwave Absorption Characteristics of Tire. In *Characterization of Minerals, Metals, and Materials*; Carpenter, J.S., Bai, C., Pablo Escobedo-Diaz, J., Hwang, J.-Y., Ikhmayies, S., Li, B., Li, J., Neves, S., Peng, Z., Zhang, M., Eds.; Springer: Cham/Basel, Switzerland, 2015; pp. 235–243.

31. Escobar-Arnanz, J.; Mekni, S.; Blanco, G.; Eljarrat, E.; Barceló, D.; Ramos, L. Characterization of organic aromatic compounds in soils affected by an uncontrolled tire landfill fire through the use of comprehensive two-dimensional gas chromatography–time-of-flight mass spectrometry. *J. Chromatogr. A* **2018**, *1536*, 163–175. [CrossRef]

32. Artíñano, B.; Gómez-Moreno, F.J.; Díaz, E.; Amato, F.; Pandolfi, M.; Alonso-Blanco, E.; Coz, E.; Garcia-Alonso, S.; Becerril-Valle, M.; Querol, X.; et al. Outdoor and indoor particle characterization from a large and uncontrolled combustion of a tire landfill. *Sci. Total Environ.* **2017**, *593–594*, 543–551. [CrossRef] [PubMed]

33. Yang, F.; Feng, W.; Liu, F.; Jing, L.; Yuan, B.; Chen, D. Experimental and numerical study of rubber concrete slabs with steel reinforcement under close-in blast loading. *Constr. Build. Mater.* **2019**, *198*, 423–436. [CrossRef]

34. Diekmann, A.; Giese, U.; Schaumann, I. Polycyclic aromatic hydrocarbons in consumer goods made from recycled rubber material: A review. *Chemosphere* **2018**, *220*, 1163–1178. [CrossRef]

35. Czajczyńska, D.; Krzyżyńska, R.; Jouhara, H.; Spencer, N. Use of pyrolytic gas from waste tire as a fuel: A review. *Energy* **2017**, *134*, 1121–1131. [CrossRef]

36. Azevedo, F.; Pacheco-Torga, F.; Jesus, C.; de Aguiar, J.B.; Camões, A.F. Properties and durability of HPC with tyre rubber wastes. *Constr. Build. Mater.* **2012**, *34*, 186–191. [CrossRef]

37. Najim Khalid, B.; Hall Matthew, R. Mechanical and dynamic properties of self-compacting crumb rubber modified concrete. *Constr. Build. Mater.* **2012**, *27*, 521–530. [CrossRef]

38. Kovler, K.; Roussel, N. Properties of fresh and hardened concrete. *Cem. Concr. Res.* **2011**, *41*, 775–792. [CrossRef]

39. Bisht, K.; Ramana, P.V. Evaluation of mechanical and durability properties of crumb rubber concrete. *Constr. Build. Mater.* **2017**, *155*, 811–817. [CrossRef]

40. Italian Organization for Standardization (UNI). Cement Composition, Specifications and Conformity Criteria for Common Cements. EN 197-1. Available online: http://store.uni.com/magento-1.4.0.1/index.php/en-197-1-2011.html (accessed on 14 September 2011).

41. Italian Organization for Standardization (UNI). Methods of Testing Cement-Part 1: Determination of Strength. EN 196-1. Available online: http://store.uni.com/magento-1.4.0.1/index.php/en-196-1-2016.html (accessed on 27 April 2016).

42. International Organization for Standardization (ISO). Cement, Test Methods, Determination of Strength. ISO 679. Available online: http://store.uni.com/magento-1.4.0.1/index.php/iso-679-2009.html (accessed on 24 April 2009).

43. Italian Organization for Standardization (UNI). Determination of Consistency of Cement Mortars Using a Flow Table. UNI 7044:1972. Available online: http://store.uni.com/magento-1.4.0.1/index.php/uni-7044-1972.html (accessed on 20 April 1972).

44. ACI Committee 544. ACI 544.2R-89. Measurement of properties of fibre reinforced concrete. In *ACI Manual of Concrete Practice, Part 5: Masonry, Precast Concrete and Special Processes*; American Concrete Institute: Farmington Hills, MI, USA, 1996.

45. Gustafsson, S.E. Transient plane source techniques for thermal conductivity and thermal diffusivity measurements of solid materials. *Rev. Sci. Instrum.* **1991**, *62*, 797–804. [CrossRef]

46. Khalil, E.; Abd-Elmohsen, M.; Anwar, A.M. Impact resistance of rubberized self-compacting concrete. *Water Sci.* **2015**, *29*, 45–53. [CrossRef]

47. Li, G.; Wang, Z.; Leung, C.K.; Tang, S.; Pan, J.; Huang, W.; Chen, E. Properties of rubberized concrete modified by using silane coupling agent and carboxylated SBR. *J. Clean. Prod.* **2016**, *112*, 797–807. [CrossRef]

48. Huang, B.S.; Li, G.Q.; Pang, S.S.; Eggers, J. Investigation into waste tire rubberfilled concrete. *J. Mater. Civ. Eng.* **2004**, *16*, 187–194. [CrossRef]

49. Khaloo, A.R.; Dehestani, M.; Rahmatabadi, P. Mechanical properties of concrete containing a high level of tire-rubber particles. *Waste Manag.* **2008**, *28*, 2472–2482. [CrossRef] [PubMed]

50. Marie, I. Thermal conductivity of hybrid recycled aggregate–Rubberized concrete. *Constr. Build. Mater.* **2017**, *133*, 516–524. [CrossRef]

51. Karakurt, C. Microstructure properties of waste tire rubber composites: An overview. *J. Mater. Cycles Waste* **2015**, *17*, 422–433. [CrossRef]

52. Di Mundo, R.; Petrella, A.; Notarnicola, M. Surface and bulk hydrophobic cement composites by tyre rubber addition. *Constr. Build. Mater.* **2018**, *172*, 176–184. [CrossRef]

53. Aiello, M.A.; Leuzzi, F. Waste tyre rubberized concrete: Properties at fresh and hardened state. *Waste Manag.* **2010**, *30*, 1696–1704. [CrossRef]

54. Topcu, I.B. The properties of rubberized concretes. *Cem. Concr. Res.* **1995**, *25*, 304–310. [CrossRef]

55. Khatib, Z.K.; Bayomy, F.M. Rubberized Portland cement concrete. *J. Mater. Civ. Eng.* **1999**, *11*, 206–213. [CrossRef]

56. Eldin, N.N.; Senouci, A.B. Rubber-tire particles as concrete aggregate. *J. Mater. Civ. Eng.* **1993**, *5*, 478–496. [CrossRef]

57. Toutanji, H.A. The use of rubber tire particles in concrete to replace mineral aggregates. *Cem. Concr. Comp.* **1996**, *18*, 135–139. [CrossRef]

58. Kou, S.C.; Poon, C.S. Properties of self-compacting concrete prepared with recycled glass aggregate. *Cem. Concr. Comp.* **2009**, *31*, 107–113. [CrossRef]

59. Ali, E.E.; Al-Tersawy, S.H. Recycled glass as a partial replacement for fine aggregate in self compacting concrete. *Constr. Build. Mater.* **2012**, *35*, 785–791. [CrossRef]

60. Petrella, A.; Petrella, M.; Boghetich, G.; Petruzzelli, D.; Ayr, U.; Stefanizzi, P.; Calabrese, D.; Pace, L. Thermo-acoustic properties of cement-waste-glass mortars. *Proc. Inst. Civ. Eng. Constr. Mater.* **2009**, *162*, 67–72. [CrossRef]

61. Petrella, A.; Spasiano, D.; Rizzi, V.; Cosma, P.; Race, M.; De Vietro, N. Lead Ion Sorption by Perlite and Reuse of the Exhausted Material in the Construction Field. *Appl. Sci.* **2018**, *8*, 1882. [CrossRef]

62. Petrella, A.; Spasiano, D.; Liuzzi, S.; Ayr, U.; Cosma, P.; Rizzi, V.; Petrella, M.; Di Mundo, R. Use of cellulose fibers from wheat straw for sustainable cement mortars. *J. Sustain. Cem. Based Mater.* **2018**, *8*, 161–179. [CrossRef]

63. Petrella, A.; Spasiano, D.; Rizzi, V.; Cosma, P.; Race, M.; De Vietro, N. Thermodynamic and kinetic investigation of heavy metals sorption in packed bed columns by recycled lignocellulosic materials from olive oil production. *Chem. Eng. Commun.* **2019**, *206*, 1715–1730. [CrossRef]

64. Petrella, A.; Cosma, P.; Rizzi, V.; De Vietro, N. Porous alumosilicate aggregate as lead ion sorbent in wastewater treatments. *Separations* **2017**, *4*, 25. [CrossRef]

65. Petrella, A.; Petrella, M.; Boghetich, G.; Petruzzelli, D.; Calabrese, D.; Stefanizzi, P.; De Napoli, D.; Guastamacchia, M. Recycled waste glass as aggregate for lightweight concrete. *Proc. Inst. Civ. Eng. Constr. Mater.* **2007**, *160*, 165–170. [CrossRef]

66. Italian Organization for Standardization (UNI). Specification on Mortar for Masonry-Mortar for Interior and Exterior Plaster. EN 998-1. Available online: http://store.uni.com/catalogo/index.php/en-998-1-2016.html (accessed on 2 September 2019).

67. Italian Organization for Standardization (UNI). Specification on Mortar for Masonry-Masonry Mortars. EN 998-2. Available online: http://store.uni.com/catalogo/index.php/uni-en-998-2-2016.html (accessed on 2 September 2019).

68. Mastali, M.; Dalvand, A.; Sattarifard, A. The impact resistance and mechanical properties of the reinforced self-compacting concrete incorporating recycled CFRP fiber with different lengths and dosages. *Compos. Part B Eng.* **2017**, *112*, 74–92. [CrossRef]

69. Palumbo, F.; Di Mundo, R. Wettability: Significance and measurement. In *Polymer Surface Characterization*; Sabbatini, L., Ed.; De Gruyter: Berlin, Germany, 2014; pp. 207–241.

Article

Nanosilica Extracted from Hexafluorosilicic Acid of Waste Fertilizer as Reinforcement Material for Natural Rubber: Preparation and Mechanical Characteristics

Van-Huy Nguyen [1,2], Cuong Manh Vu [3,*], Hyoung Jin Choi [4,*] and Bui Xuan Kien [5]

[1]   Department for Management of Science and Technology Development, Ton Duc Thang University, Ho Chi Minh City 700000, Vietnam; nguyenvanhuy@tdtu.edu.vn

[2]   Faculty of Applied Sciences, Ton Duc Thang University, Ho Chi Minh City 700000, Vietnam

[3]   Center for Advanced Chemistry, Institute of Research and Development, Duy Tan University, Da Nang 550000, Vietnam

[4]   Department of Polymer Science and Engineering, Inha University, Incheon 22212, Korea

[5]   Faculty of Natural Sciences, Electric power University, 235 Hoang Quoc Viet St., Bac Tu Liem Dist, Hanoi 100000, Vietnam

*   Correspondence: vumanhcuong@duytan.edu.vn or vumanhcuong309@gmail.com (C.M.V.); hjchoi@inha.ac.kr (H.J.C.)

Received: 30 July 2019; Accepted: 21 August 2019; Published: 23 August 2019

Abstract: Nanosilica particles are extracted from waste water containing a hexafluorosilicic acid discharged from Vietnamese fertilizer plants as an effective way not only to reduce waste water pollution but also to enhance the value of their waste water. Amorphous nanosilica particles are produced with diameters ranging from 40 to 60 nm and then adopted as a reinforcing additive for natural rubber (NR) composites. Morphological, mechanical, rheological, and thermal behaviors of the nanosilica-added NR composites are examined. Especially, mechanical behaviors of nanosilica-filled NR composites reach the optimum with 3 phr of nanosilica, at which its tensile strength, hardness, and decomposition temperature are improved by 20.6%, 7.1%, and 2.5%, respectively, compared with the pristine vulcanized NR. The improved mechanical properties can be explained by the tensile fractured surface morphology, which shows that the silica-filled NR is rougher than the pristine natural rubber sample.

Keywords: natural rubber; nanosilica; mechanical property; fertilizer plant; hexafluorosilicic acid; waste water

---

1. Introduction

While nanosilica is becoming one of the most widely used nanomaterials for many industries, with an annual growth of 5.6%, silica is used as an important filler for rubber in a range of products, such as tires and other industrial materials, because it increases its mechanical durability, heat resistance, shrinkage, thermal expansion, and stress [1–4]. In addition, it improves the wear resistance of rubber based composites by replacing a soft matrix with a hard inorganic filler. Therefore, silica-based polymer nanocomposites have many exciting features for their many applications in the automotive, electronics, marine, and other industries. Many researchers have used silica as a reinforcement material for rubber-based composites [5–9].

Mechanical characteristics of the rubber can be enhanced using nanosilica only or combined with other fillers [10,11]. In particular, high-quality silica also has an absolute advantage in the manufacture of household products (fashion footwear soles, rubber mattresses, and others) or medical rubber (gloves, boots). In the pharmaceutical industry, silica is used as a carrier for some proprietary medicines [12–14]. In the organic chemical industry, silica acts as a catalyst for some organic reactions, helps acceleration rates, and improves reaction yields [15–18]. To meet the increasing demand for silica in industrial applications, many studies focused on the fabrication of silica from different sources [19–28]. In addition to the many common sources, such as silane compounds of Na_2SiO_3, hexafluorosilicic acid (H_2SiF_6) becomes a potential economical candidate for the production of silica. Hexafluorosilicic acid is a by-product from the fertilizer industry, produced in huge quantities annually [29]. On the other hand, it is toxic and harmful to the environment, requiring either chemical treatment or conversion to a highly economical product [30]. For example, in Vietnam, Lam Thao Fertilizers and Chemicals, the largest production company of fertilizers with an estimated capacity of approximately 850,000 tons/year, produces approximately 30,000 tons of hexafluorosilicic acid annually. The production of hexafluorosilicic acid in the fertilizer production line can be explained as follows: First, fluorapatite ($Ca_5(PO_4)_3F$ (calcium fluorophosphate)) is reacted with either H_2SO_4 or HNO_3 to generate HF according to the following reactions:

$$Ca_5(PO_4)_3F + 5H_2SO_4 + nH_2O -> 3H_3PO_4 + 5CaSO_4{\cdot}nH_2O + HF$$

Or

$$Ca_5(PO_4)_3F + 10HNO_3 -> 3H_3PO_4 + 5Ca(NO_3)_2 + HF$$

Obtained HF reacts with SiO_2, which exists in the composition of the raw materials, to form SiF_4 gas:

$$4HF + SiO_2 -> SiF_4 + 2H_2O$$

The collection of H_2SiF_6 is usually performed by an absorption method of gaseous SiF_4 in a water scrubber.

$$3SiF_4 + 2H_2O = 2H_2SiF_6 + SiO_2$$

Many groups have reported its utilization, such as in silica recovery [30–37]. Dragicevic and Hraste [38] prepared silica from the neutralization of fluosilicic acid with ammonia while Sarawade et al. [25] used Na_2CO_3 to recover mesoporous silica with a large surface area from waste H_2SiF_6 from the fertilizer company. Hexafluorosilicic acid was further adopted as a silica source to fabricate SZM-5 as a trans-alkylation catalyst [29]. Cicala et al. [39] synthesized amorphous silicon alloys from fluorinated gases by plasma deposition, while Guzeev et al. [40] produced zircon and zirconium tetrafluoride with silicon tetrafluoride and zirconium dioxide as a raw material. Liu et al. [41] also synthesized titanium containing Mobil Composition of Matter No. 41 (MCM-41) with the industrial H_2SiF_6 and applied for cyclohexene epoxidation reaction.

We report a simple method of recovering amorphous silica nanoparticles from hexafluorosilicic acid waste and their application as a reinforcing filler in natural rubber (NR) with enhanced mechanical and thermal characteristics in this study. Both fabricated nanosilica and nanosilica-added NR are characterized. These efforts could not only reduce the environmental pollution of hexafluorosilicic acid wastes, but also enhance the value of waste from a fertilizer plant as an inorganic filler of NR. Note that of the total cost for the final product in factory including the cost for raw material, equipment, energies, and waste water treatment etc., the cost for waste water treatment grows higher as a result of the government policy and type of the waste water. This results in the higher price of final products, reducing their competitiveness. In case of Vietnamese fertilizer plants, the by-product of H_2SiF_6 with its emission rate is about 35,000 tons per year. With its highly toxic and corrosive characteristics, the H_2SiF_6 solution could threaten the environment by contaminating rivers and oceans.

Therefore, we strongly believe that the production and utilization of nanosilica in this study could become a reliable and sustainable solution for dealing with waste water from fertilizer plants environmentally as well as economically.

2. Experimental Procedures

2.1. Materials

The hexafluorosilicic acid was collected from Lam Thao Fertilizers and Chemicals JSC (Phu Tho, Vietnam) and used as a raw resource for the production of nanosilica. The concentration of the hexafluorosilicic acid solution was 13 wt.% along with a very low concentration of Fe_2O_3 (0.01–0.02 wt.%) and P_2O_5 (0.001–0.01 wt.%). A 25 wt.% ammonia solution was purchased from Xilong Scientific Co. (Shantou, China). The NR (NR-Standard Vietnamese Rubber-3L) was purchased from Phuoc Hoa Rubber Co. (Binh Duong, Vietnam). We adopted 2, 2, 4-trimetyl-1, 2-dihydroquinolin (RD), N-cyclohexyl-2-benzothiazole sulfenamide (CBS), and 2, 2′-dibenzothizole disulfide (DM), (RongCheng K&S Chem., Rongcheng, China) as vulcanizing accelerators [42]. Stearic acid, zinc oxide, and sulfur were purchased from Sigma-Aldrich (St. Louis, MO, USA).

2.1.1. Recovery of Nanosilica

In the first step, pristine solutions of both hexafluorosilicic acid and ammonia were diluted with distilled water to 10 wt.% and 20 wt.%, respectively. In the second step, 200 g of 10 wt.% hexafluorosilicic acid and 110 g of 20 wt.% ammonia solution were mixed in a glass reactor vessel with a mechanical stirrer at 200 rpm for 12 h at 25 °C to obtain a slurry of nanosilica. Subsequently, the silica nanoparticles were collected with a vacuum filter under atmosphere 200 mg Hg and cleaned a couple of times with distilled water until they reached a neutral pH value. The resulting products were dried using a vacuum oven at 800 °C for 3 h and cooled to 25 °C before obtaining the final product. Figure 1 provides details of this processing.

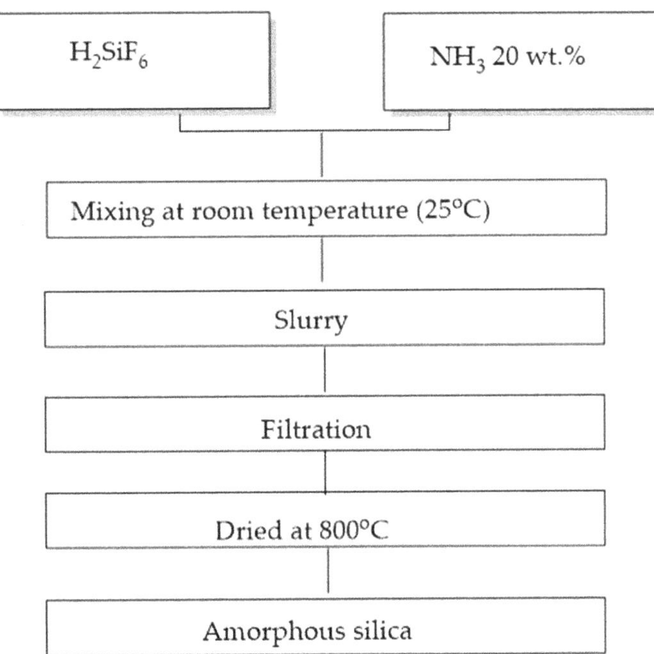

Figure 1. The processing of fabrication of nanosilica from hexafluorosilicic acid.

The material characteristics of the obtained nanosilica, such as chemical structure, crystallinity, and particle size were scrutinized by Fourier transform-infrared spectroscopy (FT-IR), X-ray diffraction (XRD), and transmission electron microscopy (TEM), respectively.

2.1.2. Rubber Compound Fabrication

The NR compounds both with and without nanosilica were fabricated following the formulations given in Table 1. Initially, the NR, ZnO, stearic acid, paraffin, and nanosilica were put into an internal mixer (Brabender Plasti-corder 350s, Brabender GmbH & Co. KG, Berlin, Germany) at different compositions depending on the sample number and mixed well at 50 °C with a rotor speed of 50 rpm for 1 h [42]. Further mixing was accomplished with a two-roll mill once three vulcanizing accelerator chemicals and sulfur were added. The rubber compounds were either used for examining curing characteristics or vulcanized to determine their mechanical properties.

Table 1. The composition of silica/natural rubber compounds.

Ingredients (phr)	M0	M1	M2	M3	M4	M5
Natural Rubber	100.0	100.0	100.0	100.0	100.0	100.0
Zinc Oxide	5.0	5.0	5.0	5.0	5.0	5.0
Stearic Acid	3.0	3.0	3.0	3.0	3.0	3.0
Parafin	1.0	1.0	1.0	1.0	1.0	1.0
RD	2.5	2.5	2.5	2.5	2.5	2.5
CBS	1.5	1.5	1.5	1.5	1.5	1.5
DM	0.5	0.5	0.5	0.5	0.5	0.5
Sulfur	2.0	2.0	2.0	2.0	2.0	2.0
Silica	0.0	1.0	3.0	5.0	7.0	10.0

2.2. Measurements

The moving die rheometer (Monsanto, MDR2000P, St. Louis, MO, USA) was used to determine the curing characteristics of NR compounds. About 5 g of rubber compound was inserted into the geometry of two parallel rotating disks at 150 °C at a frequency of 100 rpm. After processing was completed, the cure curve with many characteristics such as max. torque (M_H), min. torque (M_L), scorch time (t_2), and 90% cure time (t_{90}) were acquired.

In order to prepare the samples for tensile testing, a sheet of about 2 mm thickness was vulcanized in a molding test press (Gotech, GT7014H, Taichung, Taiwan) at 150 °C and a 40 kgf/cm^2 pressure for a respective cure time, t_{90}, which was estimated from the MDR 2000P measurement [42].

The tensile test was performed using an Instron universal testing machine according to ASTM D412-93 at room temperature (~25 °C). Tensile strengths and elongations at break were estimated from stress-strain curves and averaged values from five-time rerun measurements for each sample were obtained [34]. The Shore A hardness of the samples was evaluated following the ISO 7619-1:2010. Thermogravimetric analysis (TGA) was performed with a TGA Q50 (TA Instruments, New Castle, DE, USA) according to the ASTM D3850-94 method. Approximately 10–20 mg of vulcanized samples were loaded onto an open platinum pan, and then heated from 25 to 600 °C under a nitrogen environment at a fixed heating rate of 10 °C/min. The fracture surfaces of the fabricated nanocomposites were examined with a scanning electron microscope (SEM) (SEM JEOL 5510, JEOL, Tokyo, Japan) at 10 kV accelerating voltage.

The Mooney-Rilvin equation was used to determine the crosslinking density of the vulcanates based on the following stress-strain behavior [42]:

$$\sigma = 2(\lambda - \frac{1}{\lambda^2})(C_1 + \frac{C_2}{\lambda})$$

where σ, λ, C_1, and C_2 are the tensile stress, the strain, and constants, respectively. The C_1 and C_2 constants were determined from the slope and intercept of the curve of $\sigma/(\lambda - \lambda^{-2})$ versus $1/\lambda$. Finally, the crosslinking density was obtained from the following equation:

$$2C_1 = \rho kT$$

where ρ is the cross-linking density, k is the Boltzmann constant, and T is the absolute temperature.

3. Results and Discussion

Initially, the chemical structure of fabricated silica nanoparticles was examined with the help of an FT-IR spectroscopy (Perkin Elmer, Waltham, MA, USA) as shown in Figure 2. The silica nanoparticles exhibited a distinctive absorption peak at 1000–1100 cm^{-1}, which was ascribed to a stretching vibration of Si–O–Si bonding. Another distinctive absorption peak also appeared at 1630 cm^{-1}, which was allocated to a bending vibration by way of H–O–H in the water molecules. An additional absorption peak was further detected at 3400 cm^{-1}, which was due to a stretching vibration of the hydroxyl group, confirming the existence of hydroxyl groups in the silica surface. Figure 2 also showed the FT-IR spectrum of vulcanized natural rubber filled with 3 phr silica composite. It is seen that the asymmetrical stretching vibration of Si–O–Si and the hydroxyl of silica in the NR/SiO$_2$ composite appeared at 1088 cm^{-1} and 3403 cm^{-1}, respectively.

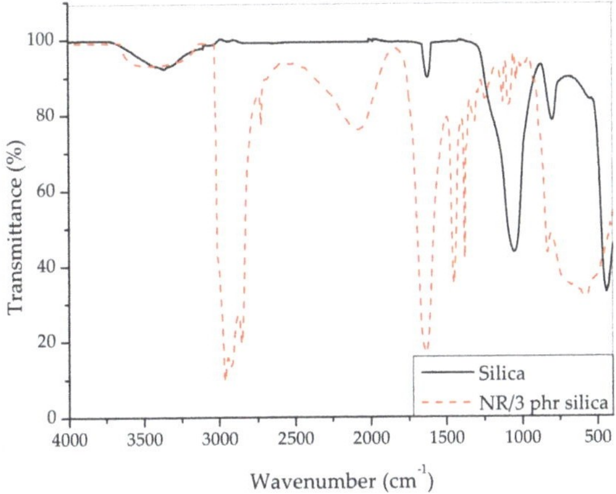

Figure 2. FT-IR spectra of fabricated nanosilica and natural rubber (NR)/3 phr silica composite.

Regarding the crystalline structure of the samples, the powder XRD pattern in Figure 3 indicated a broad peak at 22° of 2θ, revealing the amorphous nature of the nanosilica particles. Furthermore, the TEM image of nanosilica in Figure 4 showed its diameter of 40–60 nm.

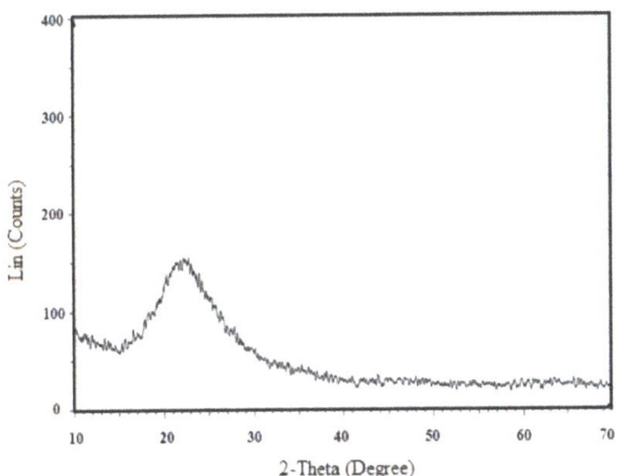

Figure 3. XRD pattern of nanosilica.

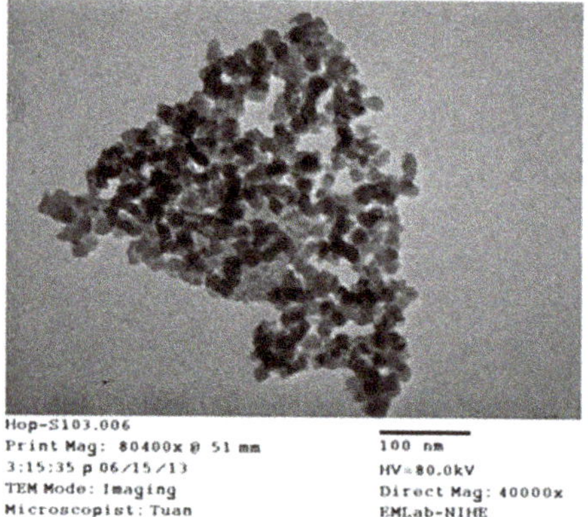

Figure 4. TEM of nanosilica.

Table 2 lists the main curing characteristics of the fabricated NR compounds that were obtained from the cure curves (Figure 5), in which both t_2 and t_{90} increased with increasing silica nanoparticle content in the NR. In general, the minimum torque (M_L) in Figure 5 is associated with a shear viscosity of the blend, while the maximum torque (M_H) has a relation to the elastic stiffness of vulcanized samples. The results implied that with addition of nanosilica, both M_L and M_H values increased, meaning that the viscosity or stiffness of vulcanized NR increased with the presence of nanosilica particles. The difference between maximum torque and minimum torque, (M_H-M_L), which was obtained from the dynamic shear modulus test, corresponds indirectly to the crosslinking density of the vulcanization. The results indicated that the (M_H-M_L) exhibited the same tendency to the maximum torque and increased with increasing nanosilica loading as a result of increased cross-linking of the NR with the presence of nanosilica particles.

Figure 6 presents typical stress-strain curves of the nanosilica-reinforced NR composite materials. All samples exhibited the deformation-forced crystallization characteristics with the rapidly raised sharp slopes when the strain reached more than 1500% [36,43,44]. Meanwhile, the tensile stress and slope curves increased with increased contents of silica and reached the optimal properties at 3 phr of silica filled NR. The details of the mechanical characteristics are provided in Table 3.

Table 2. Curing properties of the silica/rubber compounds.

Samples	Curing Properties				
	Minimum Torque, M_L (dN·m)	Maximum Torque, M_H (dN·m)	$\triangle M = M_H - M_L$ (dN·m)	Scorch Time, t_{s2} (min:s)	Cure Time, t_{90} (min:s)
M0	0.135	6.000	5.865	4:11	15:89
M1	0.144	6.400	6.256	4:25	16:15
M2	0.148	6.600	6.452	4:39	16:41
M3	0.154	6.850	6.696	4:41	16:83
M4	0.169	7.535	7.366	4:54	16:92
M5	0.185	8.220	8.035	4:60	17:01

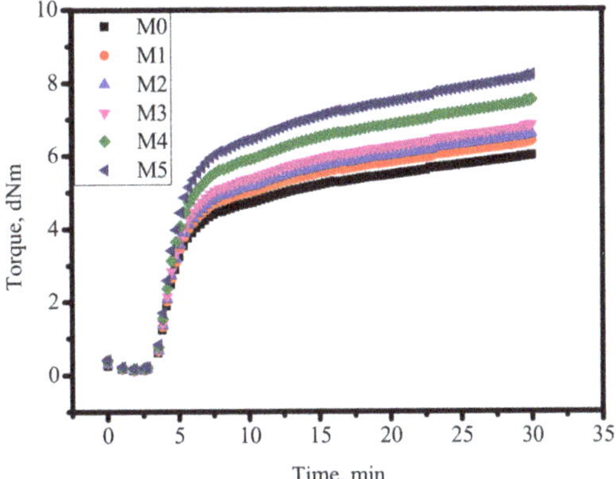

Figure 5. Cure curves of the investigated rubber compounds taken at 150 °C (black line—M0; red line—M1; blue line—M2; pink line—M3; olive line—M4; green line—M5).

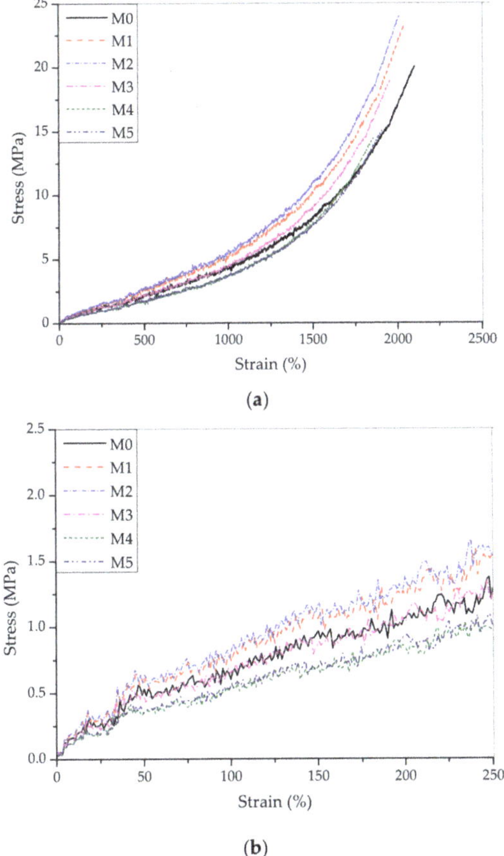

(**a**)

(**b**)

Figure 6. (**a**) Typical stress-strain of cured samples (Black line—M0; red line–M1; blue line—M2; pink line—M3; olive line—M4; orange line—M5). (**b**) The bottom figure is the one expanded in the initial stage;

Table 3. Mechanical properties of cured NR/silica compounds.

Samples	Mechanical Properties		
	Tensile Strength (MPa)	Elongation at Break (%)	Hardness (Shore A)
M0	20.02	2100.12	38.51
M1	23.18	2036.85	39.72
M2	24.15	2014.52	41.24
M3	18.93	1953.77	42.51
M4	15.13	1909.49	43.22
M5	14.45	1855.26	45.06

The tensile strength exhibited the decreasing trend when the silica content exceeded 3 phr. Although the elongation at break decreased with increased silica concentration, the hardness of the NR showed the same trend with the tensile strength. Both changes in the tensile strength and elongation at break for different nanosilica loadings have a relationship with a cross-linking density of the cured NR composites. That of the NR increased with silica concentration, resulting in an increase of the tensile strength and a decrease of the elongation at break due to the decreased slippage among the molecular chains. The H-bond between silica particles and rubber chain prevented the mobility and slippage of the rubber chain. On the other hand, the fact that tensile strength only increased with silica contents up to 3 phr may be due to the agglomeration of nanosilica at a higher content. The crosslinking densities of samples with name M0, M1, M2, M3, M4, M5 were 7.82×10^5, 8.06×10^5, 8.19×10^5, 8.22×10^5, 8.36×10^5 and 8.45×10^5 mol/cm^3, respectively. The trend of crosslinking density is the same as with the trend of M_H and hardness. The hardness also increased with increased silica content. The incremental increase of crosslinking density resulted in the decreasing of elongation at break because of the prevention of the slip between each NR molecule. In addition, more energy was needed to break the linkage between each rubber molecule as the result of increased tensile strength.

The SEM image results of the tensile fractured surface of the NR composites both in the absence and presence of 3 phr of nanosilica were presented in Figure 7. These results indicated that the fractured surface of the 3 phr silica-filled NR was observed to be rougher in comparison with the pristine sample. Therefore, extra energy was required compared to the case of the smooth fractured surface of the pristine NR. This result is in agreement with the tensile results shown in Figure 6.

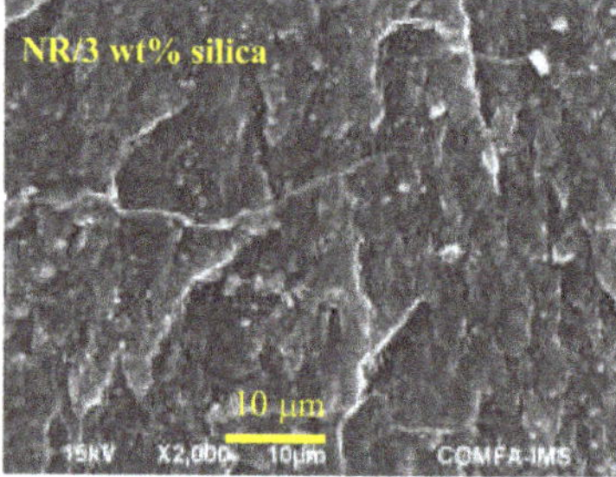

Figure 7. SEM images of the fracture surface of pristine natural rubber and 3 phr silica filled natural rubber.

The TEM images of both the natural rubber and 3 phr silica filled natural rubber are shown in Figure 8. The pristine NR sample had no silica particles, while the silica particles existed in the modified NR with nano scale from 20–60 nm.

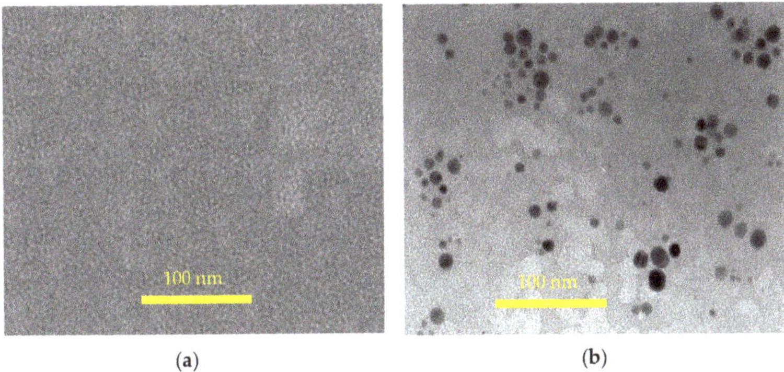

(a)   (b)

Figure 8. TEM images of pristine natural rubber (**a**) and 3 phr silica filled natural rubber (**b**).

The thermal degradation of both the pristine NR and the 3 phr nanosilica filled NR composites was examined in terms of the weight loss (%) as a function of temperature, as shown in Figure 9.

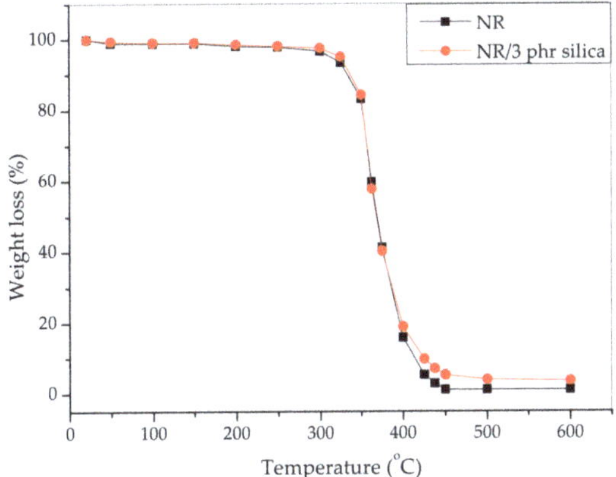

Figure 9. TGA of cured natural rubber (NR) (black dot line) and 3 phr of nanosilica filled natural rubber (red dot line).

Thermal degradation of the NR can be explained via various processes such as chain-scission of the polymers, and breakage of the cross-linked portion. The silica filled NR composite exhibited a 2.5% higher decomposition temperature in comparison with the pristine NR because of the existence of H-bonds between silica particles and rubber chains with higher thermal stability. The thermal degradation of both pristine and silica filled NR occurs through the degradation of isoprene units at around 368 °C. The presence of nanosilica induced the higher residues of nanocomposite when compared with pristine NR due to the presence of inorganic filler, which is more thermally stable than NR. The TGA results demonstrate that the nanosilica filled NR showed only a very slightly higher thermal stability than that for the NR. The initial temperature decomposition, the maximum decomposition temperature, and residue of silica modified NR and pure NR were 268.2 °C, 368.2 °C, 1.1% and 266.8 °C, 366.3 °C, 3.7%, respectively. The entire thermal degradation of the nanocomposite can be explained by the two-step process. First of all, the rubber chains and cross-linking were deteriorated into smaller parts. In the second step, the smaller parts in the first step continuously degraded into volatile products and disappeared. The residual char was higher for the nanosilica-filled sample.

Figure 10 shows the XRD patterns of vulcanized pure NR and NR/3 phr silica nanocomposite. The broad diffraction peak around 20° is the noncrystalline structure of NR, while the diffraction peaks between 30°–50° are assigned to ZnO particles in the vulcanizates. None of the samples show obvious characteristic peaks of graphite or silica, indicating that silica particles are homogeneously dispersed in the rubber matrix. These results indicated that the silica did not keep the amorphous structure when embedded in natural rubber.

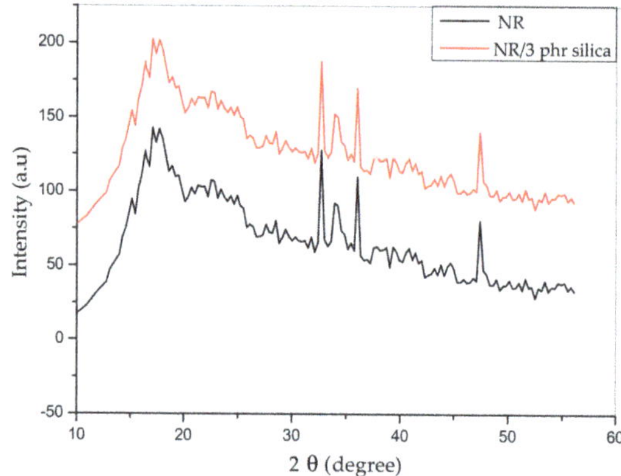

Figure 10. XRD spectra of vulcanized NR and NR/3 phr silica.

4. Conclusions

In this study, amorphous silica nanoparticles (between the sizes of 40–60 nm) were extracted from hexafluorosilicic acid waste produced by the Vietnamese fertilizer industry via a precipitation process. The production and utilization of the nanosilica by-product could become a reliable and sustainable solution for dealing with waste water from fertilizer plants environmentally as well as economically, regarding the waste water treatment. The resulting silica nanoparticles were then adopted as the filler for NR. The tensile strength of 3 phr silica-added NR nanocomposites increased by 20.6% compared to that of pristine NR. The elongation at break decreased with increased filler content and the hardness of the filled sample increased with increasing nanosilica content. The filled sample also exhibited better thermal resistance than the pristine sample due to the presence of nanosilica.

Author Contributions: Data curation, C.M.V., B.X.K.; formal analysis, C.M.V.; investigation, V.-H.N.; validation, H.J.C.; writing—original draft, C.M.V.; writing—review and editing, H.J.C.

Funding: This work was funded by Vietnam National Foundation for Science and Technology Development (grant number: 104.02-2017.15). One of the authors (H.J.C.) was partially supported by National Research Foundation of Korea (2018R1A4A1025169).

References

1. Chandra, C.S.J.; Bipinbal, P.K.; Sunil, K.N. Viscoelastic behaviour of silica filled natural rubber composites—Correlation of shear with elongational testing. *Polym. Test.* **2017**, *60*, 187–197. [CrossRef]
2. Tchalla, S.T.; Gac, P.Y.L.; Maurin, R.; Creachcade, R. Polychloroprene behaviour in a marine environment: Role of silica fillers. *Polym. Degrad. Stab.* **2017**, *139*, 28–37. [CrossRef]
3. Xu, T.; Jia, Z.; Wu, L.; Chen, Y.; Luo, Y.; Jia, D.; Peng, Z. Influence of acetone extract from natural rubber on the structure and interface interaction in NR/silica composites. *Appl. Surf. Sci.* **2017**, *423*, 43–52. [CrossRef]
4. Zhang, C.; Tang, Z.; Guo, B.; Zhang, L. Significantly improved rubber-silica interface via subtly controlling surface chemistry of silica. *Compos. Sci. Technol.* **2018**, *156*, 70–77. [CrossRef]
5. Jing, Y.; Niu, H.; Li, Y. Improved ethylene-propylene rubber/silica interface via in-situ polymerization. *Polymer* **2019**, *172*, 117–125. [CrossRef]
6. Jong, L. Improved mechanical properties of silica reinforced rubber with natural polymer. *Polym. Test.* **2019**, *79*, 106009. [CrossRef]

7. Tian, Q.; Tang, Y.; Ding, T.; Li, X.; Zhang, Z. Effect of nanosilica surface-capped by bis[3-(triethoxysilyl)propyl] tetrasulfide on the mechanical properties of styrene-butadiene rubber/butadiene rubber nanocomposites. *Compos. Commun.* **2018**, *10*, 190–193. [CrossRef]

8. Xu, H.; Xia, X.; Hussain, M.; Song, Y.; Zheng, Q. Linear and nonlinear rheological behaviors of silica filled nitrile butadiene rubber. *Polymer* **2018**, *156*, 222–227. [CrossRef]

9. Liu, D.; Song, L.; Song, H.; Chen, J.; Tian, Q.; Chen, L.; Sun, L.; Lu, A.; Huang, C.; Sun, G. Correlation between mechanical properties and microscopic structures of an optimized silica fraction in silicone rubber. *Compos. Sci. Technol.* **2018**, *165*, 373–379. [CrossRef]

10. Dong, B.; Liu, C.; Wu, Y.P. Fracture and fatigue of silica/carbon black/natural rubber composites. *Polym. Test.* **2014**, *38*, 40–45. [CrossRef]

11. Spratte, T.; Plagge, J.; Wunde, M.; Klüppel, M. Investigation of strain-induced crystallization of carbon black and silica filled natural rubber composites based on mechanical and temperature measurements. *Polymer* **2017**, *115*, 12–20. [CrossRef]

12. Cheng, Y.Z.; Zeng, X.; Cheng, D.B.; Xu, X.D.; Zhang, X.Z.; Zhuo, R.X.; He, F. Functional mesoporous silica nanoparticles (MSNs) for highly controllable drug release and synergistic therapy. *Colloids Surf. B Biointerfaces* **2016**, *145*, 217–225. [CrossRef]

13. Geng, H.; Zhao, Y.; Liu, J.; Cui, Y.; Wang, Y.; Zhao, Q.; Wang, S. Hollow mesoporous silica as a high drug loading carrier for regulation insoluble drug release. *Int. J. Pharm.* **2016**, *510*, 184–194. [CrossRef]

14. Jiao, J.; Li, X.; Zhang, S.; Liu, J.; Di, D.; Zhang, Y.; Zhao, Q.; Wang, S. Redox and pH dual-responsive PEG and chitosan-conjugated hollow mesoporous silica for controlled drug release. *Mater. Sci. Eng. C* **2016**, *67*, 26–33. [CrossRef]

15. Anbarasu, G.; Malathy, M.; Karthikeyan, P.; Rajavel, R. Silica functionalized Cu(II) acetylacetonate Schiff base complex: An efficient catalyst for the oxidative condensation reaction of benzyl alcohol with amines. *J. Solid State Chem.* **2017**, *253*, 305–312. [CrossRef]

16. Leckie, L.; Mapolie, S.F. Mesoporous silica as phase transfer agent in the biphasic oxidative cleavage of alkenes using triazole complexes of ruthenium as catalyst precursors. *Appl. Catal. A Gen.* **2018**, *565*, 76–86. [CrossRef]

17. Shabbir, S.; Lee, S.; Lim, M.; Lee, H.; Ko, H.; Lee, Y.; Rhee, H. Pd nanoparticles on reverse phase silica gel as recyclable catalyst for Suzuki-Miyaura cross coupling reaction and hydrogenation in water. *J. Organomet. Chem.* **2017**, *846*, 296–304. [CrossRef]

18. Zeng, K.; Huang, Z.; Yang, J.; Gu, Y. Silica-supported policresulen as a solid acid catalyst for organic reactions. *Chin. J. Catal.* **2015**, *36*, 1606–1613. [CrossRef]

19. Abou, R.M.; Faouzi, H. Synthesis and charaterization of amorphous silica nanoparticles from aqueous silicates using cationic Surfactants. *J. Met. Mater. Miner.* **2014**, *24*, 37–42.

20. Elineema, G.; Kim, J.K.; Hilonga, A.; Shao, G.N.; Kim, Y.N.; Quang, D.V.; Sarawade, P.B.; Kim, H.T. Quantitative recovery of high purity nanoporous silica from waste products of the phosphate fertilizer industry. *J. Ind. Eng. Chem.* **2013**, *19*, 63–67. [CrossRef]

21. Gustafsson, H.; Holmberg, K. Emulsion-based synthesis of porous silica. *Adv. Colloid Interface Sci.* **2017**, *247*, 426–434. [CrossRef]

22. Kang, K.K.; Oh, H.S.; Kim, D.Y.; Shim, G.; Lee, C.S. Synthesis of Silica Nanoparticles Using Biomimetic Mineralization with Polyallylamine Hydrochloride. *J. Colloid Interface Sci.* **2017**, *507*, 145–153. [CrossRef]

23. Kerdlap, W.; Thongpitak, C.; Keawmaungkom, S.; Warakulwit, C.; Klangprapan, S.; Choowongkomon, K.; Chisti, Y.; Hansupalak, N. Natural rubber as a template for making hollow silica spheres and their use as antibacterial agents. *Micropor. Mesopor. Mater.* **2019**, *273*, 10–18. [CrossRef]

24. Zulfiqar, U.; Subhani, T.; Husain, S.W. Synthesis and characterization of silica nanoparticles from clay. *J. Asian Ceramic Soc.* **2016**, *4*, 91–96. [CrossRef]

25. Sarawade, P.B.; Kim, J.K.; Hilonga, A.; Kim, H.T. Recovery of high surface area mesoporous silica from waste hexafluorosilicic acid (H_2SiF_6) of fertilizer industry. *J. Hazard. Mater.* **2010**, *173*, 576–580. [CrossRef]

26. Satisk, K.W. Amorphous Precipitated Siliceous Pigment for Cosmetic or Dentifrice Use and Method for Their Production. USA Patent 3,928,541, 23 December 1975.

27. Ui, S.W.; Lim, S.J.; Sang, H.L.; Choi, S.C. Control of size and morphology of nanosize silica particles using a sodium silicate solution. *J. Ceram. Process. Res.* **2009**, *10*, 553–558.

28. Patel, B.H.; Patel, P.N. Synthesis and charaterization of silica nano particles by acid leaching technique. *Res. J. Chem. Sci.* **2014**, *4*, 52–55.

29. Jin, F.; Wang, X.; Liu, T.; Wu, Y.; Xiao, L.; Yuan, M.; Fan, Y. Synthesis of ZSM-5 with the Silica Source from Industrial Hexafluorosilicic Acid as Transalkylation Catalyst. *Chin. J. Chem. Eng.* **2016**, *25*, 1303–1313. [CrossRef]

30. Krysztafkiewicz, A.; Rager, B.; Maik, M. Silica recovery from waste obtained in hydrofluoric acid and aluminum fluoride production from fluosilicic acid. *J. Hazard. Mater.* **1996**, *48*, 31–49. [CrossRef]

31. Jeong, S.Y.; Suh, J.K.; Lee, J.M.; Kwon, O.Y. Preparation of silica-based mesoporous materials from fluorosilicon compounds: Gelation of H2SiF6 in ammonia surfactant solution. *J. Colloid Interface Sci.* **1997**, *192*, 156–161. [CrossRef]

32. Panasyuk, G.P.; Azarova, L.A.; Privalov, V.L.; Belan, V.N.; Voroshilov, I.G.; Shpigun, L.K. Preparation of Silicon Dioxide with a Fixed Content of Fluorine from Fluorosilicic Acid. *Theor. Found. Chem. Eng.* **2018**, *52*, 607–611. [CrossRef]

33. Sikdar, S.K.; Moore, J.H. Recovery of Hydrofluoric Acid from Fluosilicic Acid with High pH Hydrolysis. USA Patent 4,213,951, 22 July 1980.

34. Sikdar, S.K.; Moore, J.H. Process for Producing Fluorine Compounds and Amorphous Silica. USA Patent 4,308,244, 29 December 1981.

35. Spijker, R. Process for the Preparation of Pure Silicon Dioxide. *Eur. Patent* **1983**, *94*, 139.

36. Toki, S.; Hsiao, B.S. Nature of Strain-Induced Structures in Natural and Synthetic Rubbers under Stretching. *Macromolecules* **2003**, *36*, 5915–5917. [CrossRef]

37. Zorya, L.; Krot, V. Method of high-purity silica production from hexafluorosilicic acid. *React. Kinet. Catal. Lett.* **1993**, *50*, 349–354. [CrossRef]

38. Dragicevic, T.; Hraste, M. Surface area of silica produced by neutralization of fluosilicic acid. *Chem. Biochem. Eng.* **1994**, *8*, 141–143.

39. Cicala, G.; Bruno, G.; Capezzuto, P. Plasma deposition of amorphous silicon alloys from fluorinated gases. *Pure Appl. Chem.* **1996**, *5*, 1143–1149. [CrossRef]

40. Guzeev, V.V.; D'yachenko, A.N.; Grishkov, V.N. Integrated utilization of silicon tetrafluoride and zirconium dioxide. *Russ. J. Appl. Chem.* **2003**, *76*, 1952–1955. [CrossRef]

41. Liu, T.; Jin, F.; Wang, X.; Fan, Y.; Yuan, M. Synthesis of titanium containing MCM-41 from industrial hexafluorosilicic acid as epoxidation catalyst. *Catal. Today* **2017**, *297*, 316–323. [CrossRef]

42. Vu, C.M.; Vu, H.T.; Choi, H.J. Fabrication of Natural Rubber/Epoxidized Natural Rubber/Nanosilica Nanocomposites and Their Physical Characteristics. *Macromol. Res.* **2015**, *23*, 284–290. [CrossRef]

43. Chen, L.; Guo, X.; Luo, Y.; Jia, Z.; Bai, J.; Chen, Y.; Jia, D. Effect of novel supported vulcanizing agent on the interfacial interaction and strain-induced crystallization properties of natural rubber nanocomposites. *Polymer* **2018**, *148*, 390–399. [CrossRef]

44. Chenal, J.M.; Chazeau, L.; Guy, L.; Bomal, Y.; Gauthier, C. Molecular weight between physical entanglements in natural rubber: A critical parameter during strain-induced crystallization. *Polymer* **2007**, *48*, 1042–1046. [CrossRef]

Article

Micron-Size White Bamboo Fibril-Based Silane Cellulose Aerogel: Fabrication and Oil Absorbent Characteristics

Dinh Duc Nguyen [1,2], **Cuong Manh Vu** [3,4,*], **Huong Thi Vu** [5] **and Hyoung Jin Choi** [6,*]

1 Department for Management of Science and Technology Development, Ton Duc Thang University, Ho Chi Minh City 700000, Vietnam; nguyendinhduc@tdtu.edu.vn
2 Faculty of Environment and Labour Safety, Ton Duc Thang University, Ho Chi Minh City 700000, Vietnam
3 Center for Advanced Chemistry, Institute of Research and Development, Duy Tan University, Da Nang 550000, Vietnam
4 Faculty of Chemical-Physical Engineering, Le Qui Don Technical University, 236 Hoang Quoc Viet, Ha Noi 100000, Vietnam
5 AQP research and control pharmaceuticals Joint Stock Company (AQP Pharma J.S.C) Dong Da, Ha Noi 100000, Vietnam; huongvtaqp@gmail.com
6 Department of Polymer Science and Engineering, Inha University, Incheon 22212, Korea
* Correspondence: vumanhcuong309@gmail.com (C.M.V.); hjchoi@inha.ac.kr (H.J.C.)

Received: 29 March 2019; Accepted: 25 April 2019; Published: 30 April 2019

Abstract: Micron-size white bamboo fibrils were fabricated from white bamboo and used as a source for the production of highly porous and very lightweight cellulose aerogels for use as a potential oil absorbent. The aerogels were fabricated through gelation from an aqueous alkali hydroxide/urea solution, followed by a conventional freeze-drying process. The morphology and physical properties of the aerogels were characterized by field emission scanning electron microscopy and Brunauer–Emmett–Teller surface area analysis, respectively. Successful silanization of the cellulose aerogel was confirmed by energy-dispersive X-ray spectroscopy, Fourier-transform infrared spectroscopy, and water contact angle measurements. The fabricated silane cellulose aerogel exhibited excellent absorption performance for various oil and organic solvents with typical weight gains ranging from 400% to 1200% of their own dry weight, making them promising versatile absorbents for a range of applications, including water purification.

Keywords: cellulose aerogel; oil absorbent; cellulose; white bamboo fibril; water pollution

---

1. Introduction

Water pollution caused by oil spillage related to the rapid development of the petroleum industries have serious effects on the environment and human life [1–3]. To resolve this problem, many methods have been proposed for water purification, such as water/oil separation [4–6], photocatalytic degradation [7,8], and adsorption [9–11]. Among these technologies, the use of porous absorbents with a hydrophobic surface is very promising for the rapid removal of oil from the surface of water. Various types of materials used as absorbents for cleaning oil spills have been fabricated. Oil absorbents can be sorted as inorganic minerals, synthetic organics, and natural organic materials [12]. Inorganic materials, such as fly ash and exfoliated graphite have low oil absorption capacity, whereas synthetic organic materials (i.e., polypropylene and polyurethane) possess high affinity to oil and organic solvents but cause a waste problem after their use due to their very slow degradation. Natural organic materials for oil absorption from plants (cellulose fibers) and animal residues (chitin and chitosan) have attracted increasing attention because of their renewability, low-cost, and biodegradability [13]. In addition, many kinds of natural materials, such as kapok fiber [14,15], cotton fiber [16], wool fiber [17], milkweed [18], and sawdust [19], have been exploited for the simple, effective, and inexpensive treatments of oil spills. However, most of these are hydrophilic, resulting in low oil sorption capacity. Therefore, there is still a need to find new environmental friendly absorbents with high oil absorption capacity, good selectivity, and low-cost.

Aerogels are a highly porous solid that hold up to 99% by volume of air within their pores and are known as the lowest density solid material [20]. These materials have become the most promising absorbents owing to their lightness, high porosity, and large inner surface area. Aerogels can be obtained from both inorganic sources, such as silica [21–23] and metal oxides [24], and organic sources [25–27]. Generally, the mechanism of oil sorption by aerogels is considered to be governed mainly by aerogel surface adsorption, absorption, and diffusion through the voids via interfiber capillary action [28–30], and the amount adsorbed is known to be dependent on surface area and porosity of adsorbents [31].

Concurrently, as an abundant, sustainable, renewable, and biodegradable resource of biopolymer, various cellulose-based aerogels have been investigated [32–37]. However, because the untreated cellulose aerogels have high hydrophilicity, they can absorb both oil and water during the absorption processing [36]. Therefore, the surface modification is needed to improve the hydrophobicity and absorption efficiency. The silanization processing also became one of the best ways for this aim [35,36]. Cellulose can be obtained from plants, such as bamboo. In Vietnam, bamboo is distributed widely in the north and is mainly being used as a raw material for tradition products such as toothsticks, chopsticks, floors, and some furniture. To the best of the authors' knowledge, there are few reports of the fabrication of aerogels from white bamboo and their application as both an absorbent for oil and other toxic chemicals especially at the time when water pollution by oil and chemical spills is becoming a serious problem caused by the huge development of the petrochemical industry.

In this study, bamboo was used as a source of micron-size white bamboo fibrils prior to fabricating the aerogel. The aerogels were prepared in a simple way from the gels of micron-size white bamboo fibrils (MWBFs) in an aqueous alkali hydroxide/urea solution, followed by conventional freeze-drying. The resulting aerogel was rendered hydrophobic and oleophilic after being treated with a silane compound using a common chemical vapor deposition process.

2. Experiments

2.1. Materials and Chemicals

Dendrocalamus membranaceus Munro (white bamboo, ~3 years) from Hoa Binh Province, Vietnam was used as a micron-size fiber resource. Gama-methacryloxypropyltrimethoxysilane (MEMO silane) was purchased from Evonik Industries (Ho Chi Minh, Vietnam). All other chemicals, including alkali hydroxide, urea, and ethanol, were of reagent grade (Xilong Chemical Co. Ltd, Guangdong, China). All aqueous solutions were prepared using distilled water.

2.2. Preparation of Micron-Size White Bamboo Fibrils

The MWBF was fabricated from raw white bamboo using both a steam explosion following alkaline treatment technique and the mechanical extraction method (microgrinding). Raw white bamboo (~3 years old) was first cut into bamboo culms of 50–60 cm in length using a saw machine, and placed into an autoclave with over-heated steam at 175 °C and 0.7–0.8 MPa for 60 min. The steam was then released suddenly for 5 min and the cycles of sudden steam release were repeated 9 times. Subsequently, samples were immersed in a 2% NaOH solution at 70 °C for 5 h to ensure the complete removal of the cell walls. The roller looser was then used to extract the slabs into small fibers. Finally, they were washed with fresh water until they were neutralized, and dried in an oven for 24 h at 105 °C. The resulting fibers were dispersed in water with a fiber content of 10 wt.%. They were then cut into pulp fibers using a food mixer. The pulp fibers were passed 15 times between static grind and rotating grind stones revolving at 1500 rpm (MKCA6-3, Masuko Sangyo Co. Ltd., Saitama, Japan).

The obtained MWBF (water slurry with 90% water) was treated with ethanol to remove the water and filtered using a vacuum pump to obtain a sheet of MWBF. The filtered sheet of MWBF was stirred with an additional amount of ethanol using a stirrer at 5000 rpm for 15 min. The morphology of the MWBF was examined by scanning electron microscopy (SEM) (JEOL, Tokyo, Japan), as shown in Figure 1. The SEM in Figure 1 indicates that the diameter of cellulose fiber is in range from 90 nm to 0.2 μm, but its length is in the order of tens of micron.

(A)   (B)   (C)

Figure 1. Picture of white bamboo (**A**), micron-size white bamboo fibrils (MWBFs) (**B**), and SEM image of MWBFs (**C**).

2.3. Preparation of Cellulose Aerogel

The solvent mixture of NaOH/urea/H_2O (7:12:81 w/w) was precooled to approximately 5 °C. The desired amount of MWBF samples (1.5, 2, and 2.5 wt.%) was then dispersed immediately into the solvent system under vigorous stirring at this low temperature until a semitransparent or transparent gel was achieved, depending on the MWBF concentration. At the final stage, the gel was vacuumed to remove the air bubbles. The specimen thickness was controlled to approximately 2 cm using a beaker as a mold and then immersed into ethanol to obtain the 5 wt.% solution. A 100 mL of aqueous 10 wt.% H_2SO_4 solution was then added at ambient temperature for coagulation. The resulting cellulose hydrogels were washed with excess distilled water to remove the residual chemical reagents. The sample was then frozen in a freezer at −80 °C for 24 h and freeze-dried using a FTS Systems Dura-Stop Digital Control Stoppering Tray Dryer to obtain the desired cellulose aerogel (CA).

2.4. Fabrication of Silane Treated Cellulose Aerogel

A thermal chemical vapor deposition technique was used for surface modification of the cellulose aerogel. A Petri dish containing MEMO silane was placed in a vacuum desiccator together with the aerogel samples. The desiccator was sealed and vacuumed to 0.01 MPa, and then heated in an oven at 110 °C for various periods of time to determine the optimal conditions for the silanization reaction. Subsequently, the silane-coated cellulose aerogel (SCA) sample was kept in a vacuum oven for 30 min to remove the excess unreacted silane and by-products.

2.5. Characterization

All the tests were carried out in triplicate and the average results are reported. Initially, the densities of the cellulose aerogels were calculated by measuring the mass and volume of the aerogels. The mass was measured using an analytical balance, Fisher Scientific Accu-225D, with accuracy of 0.1 mg. The volume was determined by measuring the dimensions using a digital Vernier caliper. Average density was estimated after 5 measurements for 3 different aerogels [26]. The porosity was calculated using Equation (1):

$$\text{Porosity } (\%) = \left(1 - \frac{\rho_a}{\rho_c}\right) \times 100 \tag{1}$$

where ρ_a is the density of the aerogel and ρ_c is the density of MWBF (1.59 g/cm^3).

The BET specific surface area was determined by a N$_2$ physisorption method using Gemini VII 2390 equipment (Micromeritic Instrument Co., Norcross, GA, USA). The wettability of the SCA was evaluated by measuring the water contact angle. Images of distilled droplets on the SCA surface were taken with a digital camera (Cannon 20D and Nikkon 105 mm 1:2.5 lens, Bangkok, Thailand) and imported into the measurement software. The volume of the droplet was fixed using a 5-mL cylinder. The software is licensed image processing and analysis in Java (ImageJ) [38] and included low-bond axisymmetric drop shape analysis (LB-ADSA). The mean of the three measurements performed at different surface locations are reported as the water contact angle. The Fourier-transform infrared (FTIR) spectra of the cellulose aerogel (CA) sample before and after silanization were recorded on IRAFFINITY-1S equipment (Shimadzu, Kyoto, Japan) at room temperature. The microstructure and elemental analysis of the uncoated and coated CA were examined by field-emission scanning electron microscopy (FE-SEM) (JEOL JSM-7600F, Tokyo, Japan) equipped with an energy-dispersive X-ray spectroscope. The sample was coated with a thin layer of platinum by sputtering.

To investigate its compressive properties, a cylinder sample with a diameter of 20 mm and a height of 13 mm was compressed to 80% of its original height by using a universal testing machine (Instron, 100 kN, Norwood, MA, USA) with a compressing speed of 10 mm. min^{-1}. Five samples were tested to calculate the average value.

2.6. Oil/Solvent Absorption Capacity Measurements

The absorption capacity of SCA for various oils and organic solvents was determined by dipping a piece of SCA directly into the liquid (oil or solvent) for a certain time. The wet sample was then removed from the liquid and weighed after the aerogel surface has been blotted with filter paper to remove the excess surface oil/solvent. The test was repeated several times until the absorption process reached equilibrium. The absorption capacity (Q) was calculated from the mass gain using

$$Q\ (\%) = \frac{W - W_0}{W_0} \times 100 \tag{2}$$

where W_0 and W are the weights of the SCA before and after absorption, respectively.

The pseudo-first-order model (Equation (3)) and pseudo-second-order model (Equation (4)) were used to evaluate the absorption kinetics, where k_1 (h^{-1}) and k_2 (g·g^{-1}·%$^{-1}$·h^{-1}) are the adsorption rate constants of the pseudo-first-order equation and the pseudo-second-order equation, respectively. In addition, both Q_m and Q_t are absorption capacities at equilibrium conditions and at time t, respectively.

$$\ln \frac{Q_m}{Q_m - Q_t} = k_1 t \tag{3}$$

$$\frac{t}{Q_t} = \frac{1}{Q_m} t + \frac{1}{k_2 Q_m^2} \tag{4}$$

To examine their reusability, the oil/organic swollen samples were squeezed by hand to remove the absorbed solvent. The weights of the aerogels before organic adsorption, after adsorption, and after squeeze for removal of organic were measured during each cycle. Five samples were tested for each experiment.

3. Results and Discussion

3.1. Cellulose Aerogel Characteristics

By altering the concentration of the MWBF dispersion from 1.5 to 2.5 wt.%, aerogels with different porosities were prepared using a freeze-drying method. Under freeze-drying conditions, a slight shrinkage was observed in these aerogels compared to their initial hydrogel dimensions.

The experiments showed that the dispersion with 2.5 wt.% MWBF had very high viscosity, making it difficult to remove the air bubbles, resulting in a poor physical property. Table 1 lists the physical properties of the obtained cellulose aerogel.

Table 1. Physical properties of cellulose aerogels.

Sample	Cellulose Content (%)	Coagulation Solution	Density (g/cm^3)	Porosity (%)	Surface Area (m^2/g)
Cellulose aerogel	1.5	Ethanol	0.085	94.46	–
		Sulfuric acid	0.116	92.42	13.419
	2	Ethanol	0.131	91.41	–
		Sulfuric acid	0.135	91.19	9.046
	2.5	Ethanol	0.144	90.58	8.155

As shown in Table 1, the densities of the cellulose aerogels ranged from 0.085 to 0.144 g/cm^3, and all cellulose aerogels exhibited very high porosity (>90%). On the other hand, the specific surface area exhibited the low values from 8.155 to 13.419 m^2/g. Actually, this is quite common for aerogels obtained through a freeze-drying process. The main point here is that the typical porosity of such system is on the macron-scale, and hence the resulting specific surface area value is modest. The similar trend was also reported by Wang et al. [39]. Figure 2 shows the microstructure of a cross-section of the cellulose aerogel at different magnifications. The cellulose aerogels possessed a highly open porous honeycomb-like structure with a pore size distribution varying over a wide range from several to tens of micrometers. In addition, a network of interconnected uniform cellulose fibers appeared on the surface of the pore wall.

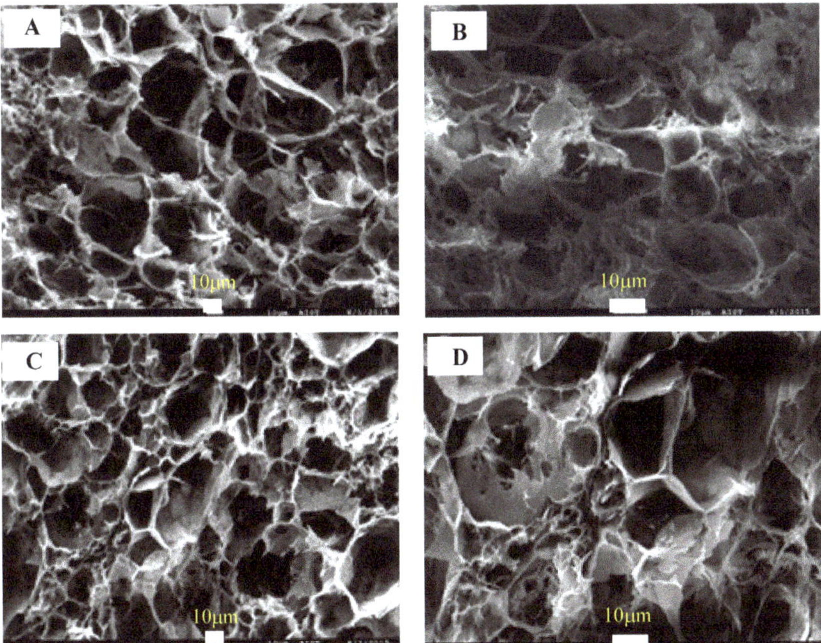

Figure 2. Field-emission scanning electron microscopy (FE-SEM) images of cellulose aerogel 2 wt.% (**a,b**) and cellulose aerogel 1.5 wt.% (**c,d**).

3.2. Mechanical Properties of Aerogel

The typical compression stress–strain curve of CA with different cellulose content is shown in Figure 3.

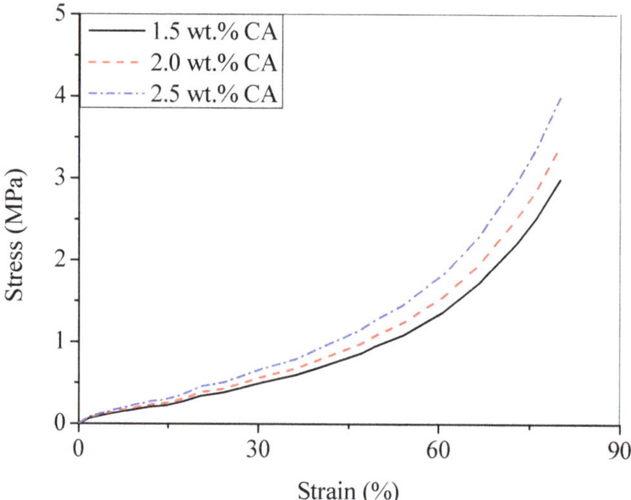

Figure 3. Typical compression stress–strain curves of cellulose aerogel (CA) with various cellulose concentration.

Figure 3 could be explained by two stress regions with increasing cellulose content. The first region appeared before 60% strain and can be characterized by slowly increasing stress, while the second region appeared at above 60% strain with rapidly increasing stress. The stress of CA obtained from 1.5, 2.0, and 2.5 wt.% of cellulose concentration at 60% strain was 1.37 ± 0.01; 1.55 ± 0.01, and 1.83 ± 0.02 MPa, respectively. At 80% strain, the stress of corresponding cellulose aerogels increased to 2.99 ± 0.02, 3.41 ± 0.01, and 4.01 ± 0.02 MPa, respectively. These results mean that the stress increased with increasing cellulose concentration as a result of higher crosslinking density. In addition, the slope of stress–strain curves was also increased with increasing cellulose content in the CA. The slope of stress–strain corresponded to the compressive modulus and the internal structure of aerogel. The high crosslinking density also was considered as a reason of increment of the compressive modulus above.

3.3. Silane Modification of Cellulose Aerogel

As hydrophilicity is the inherent nature of cellulose due to its hydroxyl groups, a silane coating was carried out for cellulose aerogels to make them both hydrophobic and oleophilic. To achieve this, a simple thermal chemical vapor deposition procedure was performed for the aerogel with MEMO silane at 110 °C in a vacuum desiccator, as described in the above. To become "active", the silane must first be hydrolyzed. The reaction of the silicon end of the molecule was initiated by the hydrolysis of the alkoxy group, usually after exposure to ambient moisture to form a silanol that releases alcohol as follows:

$$-\text{Si(OCH}_3)_3 + \text{H}_2\text{O} \rightarrow -\text{Si(OH)}_3 + 3\,\text{CH}_3\text{OH}$$

Once in the silanol state, the silane can be condensed on the aerogel surface, forming a direct covalent bond with the surface. The silanization process of the interior surface of as-prepared cellulose aerogel and silanized cellulose aerogel was shown in Figure 4.

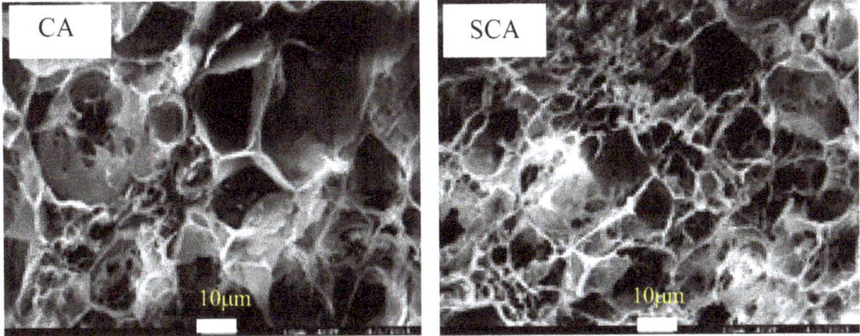

Figure 4. Silanization processing.

As shown in Figure 5, the open porous microstructure of the cellulose aerogel was preserved after coating. In addition, there was no change in the surface of the cellulose aerogel after coating, which might be due to the very thin silane coating layer.

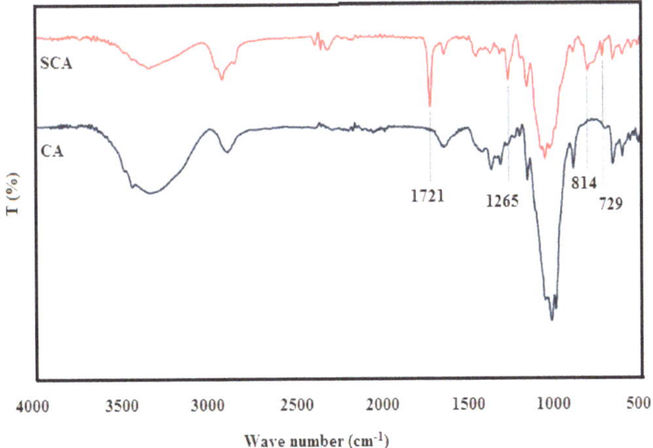

Figure 5. FE-SEM images of cross-section of CA and silane-coated cellulose aerogel (SCA).

Successful silanization on the surface of the cellulose aerogel was confirmed by FTIR spectroscopy, as shown in Figure 6. The FTIR spectrum of SCA showed four new peaks compared to the spectrum of the uncoated aerogel. The absorption bands at ~731 cm^{-1} and ~1269 cm^{-1} were attributed to the stretching and bending vibrations of the C–Si linkage, respectively. This confirmed the condensation of silane on the CA surface. In addition, the peak at 1720 cm^{-1} on the SCA spectrum was assigned to the characteristic vibrations of the carbonyl group of MEMO silane attached to the aerogel surface. The absorption band at ~815 cm^{-1} was attributed to the vibration of the Si-O-Si linkage, which might have formed due to the self-condensation of the silanols.

Figure 6. FTIR spectrums of cellulose aerogel and cellulose aerogel (CA) silanization with MEMO silane (SCA).

Silanization was confirmed by energy dispersive X-ray spectroscopy (EDX). The EDX spectrum of the uncoated CA revealed carbon and oxygen peaks but no silicon peak. After silanization, EDX showed peaks for carbon, oxygen, and silicon, as shown in Figure 7.

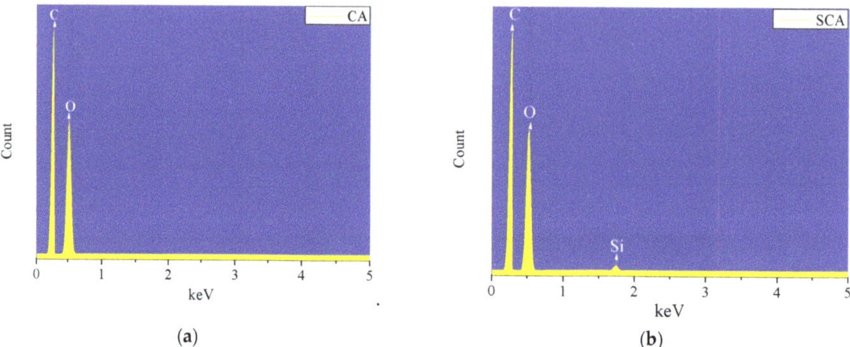

Figure 7. Energy dispersive X-ray spectroscopy (EDX) spectrum of (**a**) cellulose aerogel (CA) and (**b**) cellulose aerogel silanization with MEMO silane (SCA).

3.4. Surface Wettability of Silane-coated cellulose aerogel

Water contact angle measurements were also carried out on the uncoated and coated aerogel to study their surface wettability. The uncoated CA sample absorbed the distilled water droplet immediately in the test so no measurable contact angle was recorded. In contrast, high contact angles are observed for the MEMO silane-coated aerogel, as shown in Figure 8, proving the hydrophobicity of the material.

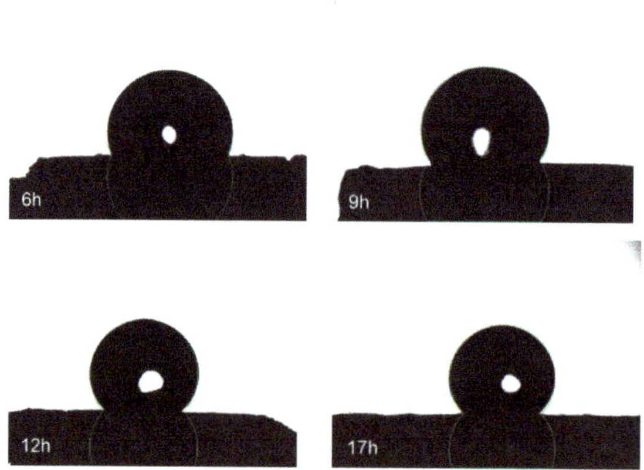

Figure 8. Measurements of water contact angle of cellulose aerogels after different coating processes.

Table 2 shows the improvement of the water contact angle of cellulose aerogels from 114.1 ± 5.26° to 132.3 ± 4.68° when the silanization time was increased from 6 to 12 h, indicating the high hydrophobicity of the obtained SCA. No increase in water contact angle was observed with further extended reaction times.

Table 2. Change of coated cellulose aerogel's water contact angle with silianation time.

Silanization Time (h)	Contact Angle (°)
6	114.1 ± 5.26
9	125.5 ± 4.68
12	132.3 ± 6.92
17	131.8 ± 5.18

3.5. Oil/Solvent Absorption Capacity of SCA and Its Recycleability

Owing to their low density, high porosity, and surface hydrophobicity, the silane-treated cellulose aerogels may be an ideal candidate for the selective absorption of oils and organic solvents from water. To examine the oil/solvent absorption behavior of MEMO silane-coated cellulose aerogel, several oils and organic solvents, such as toluene and gasoline were used.

Figure 9 shows the first minutes of the waste motor oil sorption process. The material absorbed the motor oil easily while floating in water, indicating high capacity absorption of the aerogel. After 5 min, there was no trace of waste motor oil on the water, showing that the sorption was completed successfully. Figure 10 shows the sorption kinetics of the oils and solvents on the silane-coated cellulose aerogel. The absorption rates were quite high at the very first stage and saturation was achieved after 75 h for all types of oils and solvents.

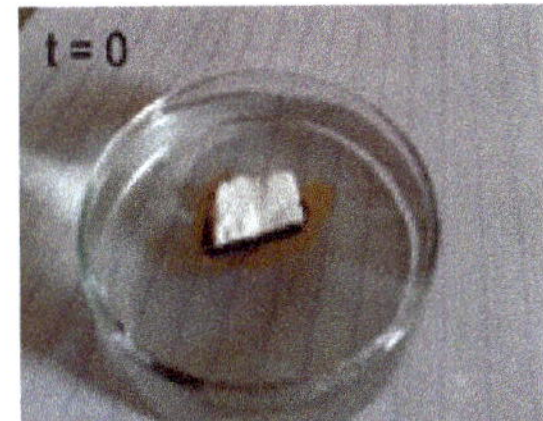

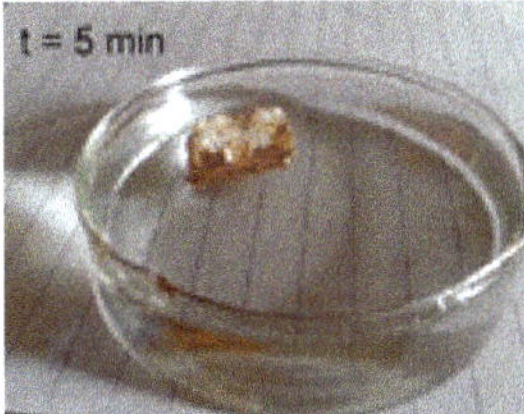

Figure 9. Waste motor oil absorption test of the silanized cellulose aerogel with MEMO silane.

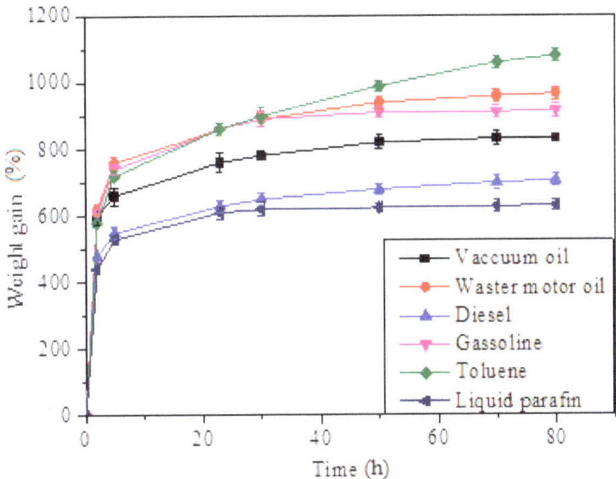

Figure 10. Absorption kinetics of oils and organic solvents on the silane-coated cellulose aerogel.

The experiment showed the linear relationship for both $\ln(Q_m/(Q_m - Q_t))$ and t/Q_t versus absorption time, t, for two representative adsorbates: vacuum oil and waste motor oil as shown in Figure 11. These results mean that the adsorption kinetics of the SCA followed the pseudo-first-order and pseudo-first-order equation quite similarly.

Equations (3) and (4) were used to calculate the absorption rate constants k_1, k_2, and correlation coefficient R^2 from Figure 11 as seen in Table 3.

The results in Table 3 indicated that the R^2 value of the pseudo second-order model of vacuum oil is higher than that of the pseudo first-order model. While the R^2 value of the pseudo second-order model of waste motor oil is lower than that of the pseudo first-order model. These results mean that the pseudo second order model can predict better the oil absorption behavior for vacuum oil and the pseudo first order model is better for waste motor oil in this work. The absorption processing of vacuum oil is faster than the absorption of waster motor oil because the k_1 and k_2 values of vacuum oil is higher than those of waste motor oil.

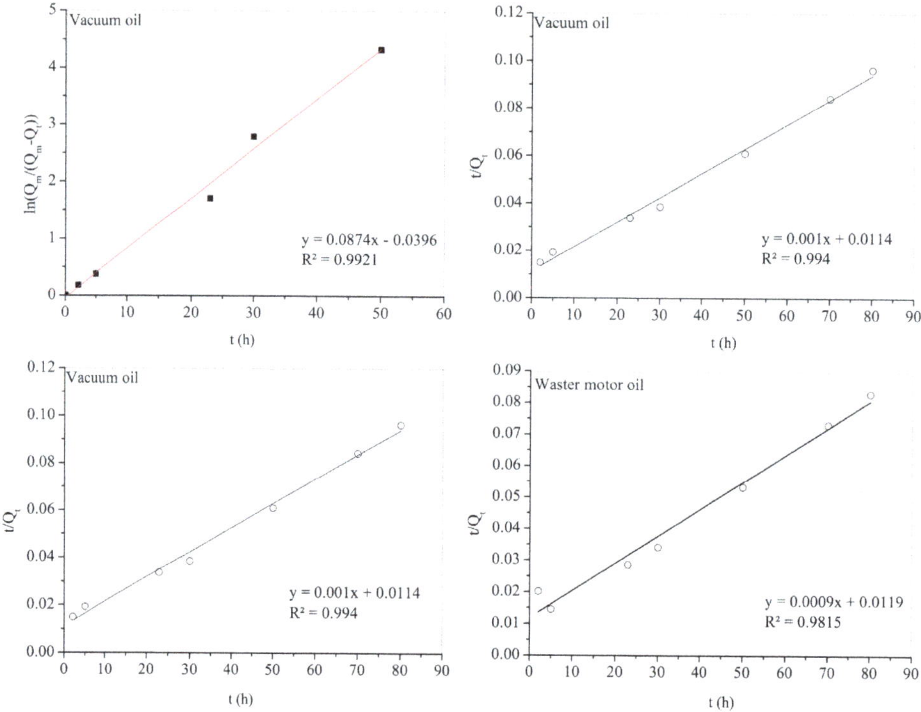

Figure 11. Pseudo-first-order and pseudo-second-order absorption linear fitting of the vacuum oil and waste motor oil onto SCA.

Table 3. Summary of the maximum oil absorption capacities and the absorption rate constants of the SCA using the pseudo-first-order and pseudo-second-order models.

Adsorbate	Maximum Absorption Capacity	Pseudo-First-Order		Pseudo-Second-Order	
	Q_m (%)	k_1	R^2	k_2	R^2
Vacuum oil	831 ± 12.2	0.0874	0.9921	1.27×10^{-5}	0.994
Waste motor oil	968 ± 18.6	0.0707	0.9956	8.99×10^{-5}	0.9815

Figure 12 shows the maximum absorption capacities of the oils and organic solvents on the silane-coated cellulose aerogel. The results showed that the absorbent had sorption capacity ranging from 631 ± 15.9% to 1081 ± 20.1% by weight gain. The high oil/solvent absorption capability of the silane-coated cellulose aerogel can be attributed to its highly porous structure and hydrophobic silane coating.

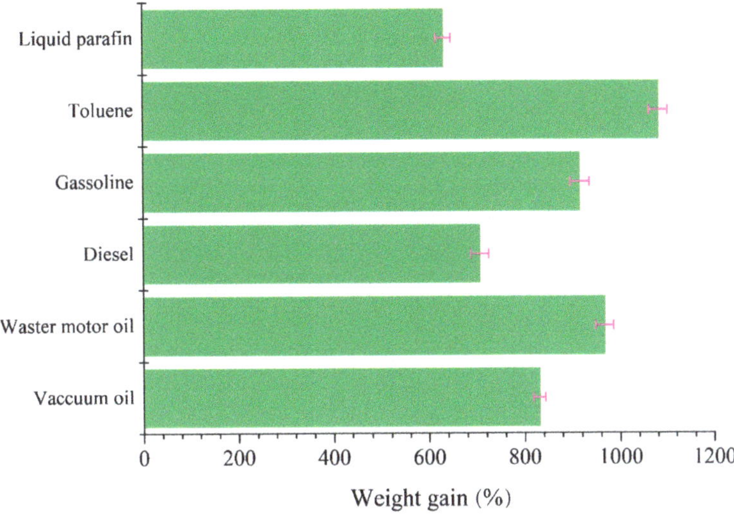

Figure 12. Absorption capacities of silane-coated cellulose aerogel for various oils and organic solvents as indicated by weight gain.

Because the weight gain of aerogel is related to the density of the respective oils and organic solvents, it can be normalized by dividing the oils and organic weight gain by the density of each respective oil and organic solvent. The results are reported in Figure 13. As shown in Figure 13, the highest absorption capacity was found for toluene and gasoline probably because these organic solvents possess the lowest viscosity. A lower viscosity would facilitate the penetration of solvent into the porous network of the aerogel more easily, leading to a higher oil/solvent absorption capacity.

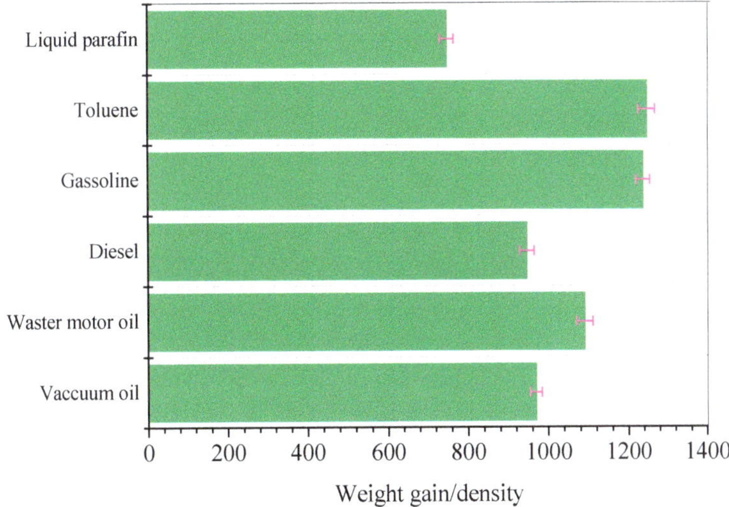

Figure 13. Absorption capacities normalized by the density of the respective oil or organic solvent.

Furthermore, for the recyclability test, the used absorbent was directly squeezed by hand and reused to absorb the oil and organic solvent. The absorption capacities of SCA for ten cycles are shown in Figure 14.

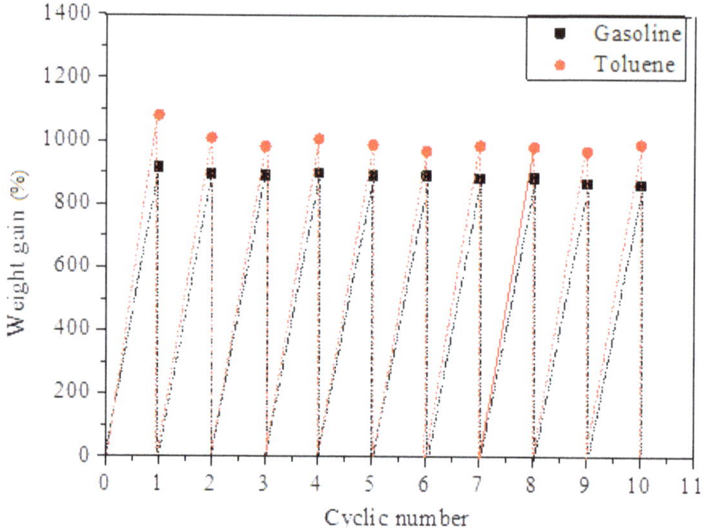

Figure 14. The cyclic adsorption capacity of sample SCA for gasoline and toluene.

After ten cycles, the adsorption capacity of the SCA for representative gasoline decreased from 916% to 862%. For toluene, the adsorption capacity decreased from 1081% to 989 % after ten cycles.

4. Conclusions

This paper reported the fabrication of a low density (0.084 g/cm^3) and highly porous (94.5%) green aerogel for the cleaning of oil and organic solvents from micron-size white bamboo fibrils (MWBF) with a very simple alkaline/ urea mixture solution method followed by a common freeze-drying process. FT-IR and EDX characterization were used to examine the surface morphology and chemical compositions of the silane-modified cellulose aerogel. The coating with MEMO silane for oil absorption purposes made the cellulose aerogel highly hydrophobic with water contact angles larger than 132.3 ± 6.92° and exhibited high absorption capacities of 1091 ± 19.6%, 1237 ± 17.6%, and 1247 ± 21.1% by weight gain for waste motor oil, diesel, and gasoline, respectively. Based on these results, the modified aerogels can be used to clean up oil spills and toxic chemicals in aquatic environments with the recyclability over 10 times.

Author Contributions: Conceptualization: D.D.N.; methodology: C.M.V.; validation: H.V.T.; formal analysis: C.M.V. and H.J.C.; investigation: C.M.V.; data curation: C.M.V. and D.D.N.; writing—original draft preparation: C.M.V. and H.J.C.; writing—review and editing: C.M.V. and H.J.C.

Funding: This research was funded by the Vietnam National Foundation for Science and Technology Development (NAFOSTED) under grant number 104.02-2017.15 (For Cuong Manh Vu) and HJC was partially supported by the National Research Foundation, Korea (2018R1A4A1025169).

Conflicts of Interest: The authors declare no conflict of interest.

References

1. Zhang, C.; Zhu, P.C.; Tan, L.; Liu, J.M.; Tan, B.; Yang, X.L.; Xu, H.B. Triptycene-Based Hyper-Cross-Linked Polymer Sponge for Gas Storage and Water Treatment. *Macromolecules* **2015**, *48*, 8509–8514. [CrossRef]
2. Howarth, A.J.; Liu, Y.; Hupp, J.T.; Farha, O.K. Metal-organic frameworks for applications in remediation of oxyanion/cation-contaminated water. *Cryst. Eng. Comm.* **2015**, *17*, 7245. [CrossRef]
3. Rey, A.; Mena, E.; Ch´avez, A.M.; Beltran, F.J.; Medina, F. Influence of structural properties on the activity of WO$_3$catalysts for visible light photocatalytic ozonation. *Chem. Eng. Sci.* **2015**, *126*, 80–90. [CrossRef]

4. Du, X.; You, S.; Wang, X.; Wang, Q.; Lu, J. Switchable and simultaneous oil/water separation induced by prewetting with a superamphiphilic self-cleaning mesh. *Chem. Eng. J.* **2017**, *313*, 398–403. [CrossRef]

5. Luo, Z.Y.; Lyu, S.S.; Fu, Y.X.; Heng, Y.; Mo, D.C. The Janus effect on superhydrophilic Cu mesh decorated with Ni-NiO/Ni(OH)$_2$ core-shell nanoparticles for oil/water separation. *Appl. Sur. Sci.* **2017**, *409*, 431–437. [CrossRef]

6. Liu, L.; Lei, J.; Li, L.; Zhang, R.; Mi, N.; Chen, H.; Huang, D.; Li, N. A facile method to fabricate the superhydrophobic magnetic sponge for oil-water separation. *Mater. Lett.* **2017**, *195*, 66–70. [CrossRef]

7. Ramezanalizadeh, H.; Manteghi, F. Design and development of a novel BiFeO$_3$/CuWO$_4$ heterojunction with enhanced photocatalytic performance for the degradation of organic dyes. *J. Photochem. Photobio A Chem.* **2017**, *338*, 60–71. [CrossRef]

8. Zhou, P.; Xie, Y.; Fang, J.; Ling, Y.; Yu, C.; Liu, X.; Dai, Y.; Qin, Y.; Zhou, D. CdS quantum dots confined in mesoporous TiO$_2$ with exceptional photocatalytic performance for degradation of organic pollutants. *Chemosphere* **2017**, *178*, 1–10. [CrossRef] [PubMed]

9. Kunaseth, M.; Poldorn, P.; Junkeaw, A.; Meeprasert, J.; Rungnim, C.; Namuangruk, S.; Kungwan, N.; Inntam, C. A DFT study of volatile organic compounds adsorption on transition metal deposited graphene. *Appl. Surf. Sci.* **2017**, *396*, 1712–1718. [CrossRef]

10. Yu, X.; Sun, W.; Ni, J. LSER model for organic compounds adsorption by single-walled carbon nanotubes: Comparison with multi-walled carbon nanotubes and activated carbon. *Environ. Pollul.* **2015**, *206*, 652–660. [CrossRef]

11. Doshi, B.; Repo, E.; Heiskanen, J.P.; Sirviö, J.A.; Sillanpää, M. Effectiveness of N,O-carboxymethyl chitosan on destabilization of Marine Diesel, Diesel and Marine-2T oil for oil spill treatment. *Carbohy. Polym.* **2017**, *167*, 326–336. [CrossRef] [PubMed]

12. Adebajo, M.O.; Frost, R.L.; Kloprogge, J.T.; Carmody, O.; Kokot, S. Porous Materials for Oil Spill Cleanup: A Review of Synthesis and Absorbing Properties. *J. Porous Mater.* **2003**, *10*, 159–170. [CrossRef]

13. Wahi, R.; Chuah, L.A.; Choong, T.S.; Ngaini, Z.; Nourouzi, M.M. Oil removal from aqueous state by natural fibrous sorbent: An overview. *Sep. Purif. Technol.* **2013**, *113*, 51–63. [CrossRef]

14. Huang, X.F.; Lim, T.T. Perfomance and mechanism of a hydrophobic-oleophilic kapok filter for oil/water separation. *Desalination* **2006**, *190*, 295–307. [CrossRef]

15. Wang, J.; Zheng, Y.; Wang, A. Investigation of acetylated kapok fibers on the sorption of oil in water. *J. Environ. Sci.* **2012**, *25*, 246–253. [CrossRef]

16. Deschamps, G.; Caruel, H.; Borredon, M.E.; Bonnin, C.; Vignoles, C. Oil Removal from water by selective sorption on hydrophobic cotton fibers. 1. Study of sorption properties and comparison with other cotton fiber-based sorbents. *Environ. Sci. Tech.* **2003**, *37*, 1013–1015. [CrossRef]

17. Rajakovic, V.; Aleksic, G.; Radetic, M.; Rajakovic, L. Efficiency of oil removal from real wastewater with different sorbent materials. *J. Hazar. Mater.* **2007**, *143*, 494–499. [CrossRef] [PubMed]

18. Choi, H.M.; Cloud, R.M. Natural sorbents in oil spill cleanup. *Environ. Sci. Tech.* **1992**, *26*, 772–776. [CrossRef]

19. Annunciado, T.R.; Sydenstricker, T.H.D.; Amico, S.C. Experimental investigation of various vegetable fibers as sorbent materials for oil spills. *Marin. Pollut. Bull.* **2005**, *50*, 1340–1346. [CrossRef]

20. Aegerter, M.A.; Leventis, N.; Koebel, M.M. *Aerogel Handbook*; Springer: New York, NY, USA, 2011; pp. 3–18.

21. Fang, W.Z.; Zhang, H.; Chen, L.; Tao, W.Q. Numerical predictions of thermal conductivities for the silica aerogel and its composites. *Appl. Therm. Eng.* **2017**, *115*, 1277–1286. [CrossRef]

22. Nazeran, N.; Moghaddas, J. Synthesis and characterization of silica aerogel reinforced rigid polyurethane foam for thermal insulation application. *J. Non-Crystall. Solid* **2017**, *461*, 1–11. [CrossRef]

23. Strobach, E.; Bhatia, B.; Yang, S.; Zhao, L.; Wang, E.N. High temperature annealing for structural optimization of silica aerogels in solar thermal applications. *J. Non-Crystall. Solid* **2017**, *462*, 72–77. [CrossRef]

24. Wu, X.; Li, W.; Shao, G.; Shen, X.; Cui, S.; Zhou, J.; Wei, Y.; Chen, X. Investigation on textural and structural evolution of the novel crack-free equimolar Al$_2$O$_3$-SiO$_2$-TiO$_2$ ternary aerogel during thermal treatment. *Ceramic. Inter.* **2017**, *43*, 4188–4196. [CrossRef]

25. Zheng, Q.; Cai, Z.; Gong, S. Green synthesis of polyvinyl alcohol (PVA)–cellulose nanofibril (CNF) hybrid aerogels and their use as superabsorbents. *J. Mater. Chem. A* **2014**, *2*, 3110–3118. [CrossRef]

26. Tripathi, A.; Parsons, G.N.; Khan, S.A.; Rojas, O.J. Synthesis of organic aerogels with tailorable morphology and strength by controlled solvent swelling following Hansen solubility. *Sci. Rep.* **2018**, *8*, 2106. [CrossRef]

27. Guo, H.; Meador, M.A.B.; McCorkle, L.; Quade, D.J.; Guo, J.; Hamilton, B.; Cakmak, M.; Sprowl, G. Polyimide Aerogels Cross-Linked through Amine Functionalized Polyoligomeric Silsesquioxane. *ACS Appl. Mater. Interfaces* **2011**, *3*, 546–552. [CrossRef]

28. Feng, J.; Wang, X.; Jiang, Y.; Du, D.; Feng, J. Study on Thermal Conductivities of Aromatic Polyimide Aerogels. *ACS Appl. Mater. Interfaces* **2016**, *8*, 12992–12996. [CrossRef]

29. Zhang, Y.; Yin, M.; Lin, X.; Ren, X.; Huang, T.S.; Kim, I.S. Functional Nanocomposite Aerogels Based on Nanocrystalline Cellulose for Selective Oil/Water Separation and Antibacterial Applications. *Chem. Eng. J.* **2019**, *371*, 306–313. [CrossRef]

30. Yue, X.; Zhang, T.; Yang, D.; Qiu, F.; Li, Z. Hybrid aerogels derived from banana peel and waste paper for efficient oil absorption and emulsion separation. *J. Clean. Prod.* **2018**, *199*, 411–419. [CrossRef]

31. Carmody, O.; Frost, R.; Xi, Y.; Kokot, S. Surface characterisation of selected sorbent materials for common hydrocarbon fuels. *Surf. Sci.* **2007**, *601*, 2066–2076. [CrossRef]

32. Silva, T.C.F.; Habibi, Y.; Colodette, J.L.; Elder, T.; Lucia, L.A. A fundamental investigation of the microarchitecture and mechanical properties of tempo-oxidized nanofibrillated cellulose (NFC)-based aerogels. *Cellulose* **2012**, *19*, 1945–1956. [CrossRef]

33. Sai, H.; Xing, L.; Xiang, J.; Cui, L.; Jiao, J.; Zhao, C.; Lia, Z.; Lia, F. Flexible aerogels based on an interpenetrating network of bacterial cellulose and silica by a non-supercritical drying process. *J. Mater. Chem. A* **2013**, *1*, 7963–7970. [CrossRef]

34. Cai, J.; Kimura, S.; Wada, M.; Kuga, S.; Zhang, L. Cellulose Aerogels from Aqueous Alkali Hydroxide–Urea Solution. *Chem. Sus. Chem.* **2008**, *1*, 149–154. [CrossRef]

35. Cervin, N.T.; Aulin, C.; Larsson, P.T.; Wagberg, L. Ultra porous nanocellulose aerogels as separation medium for mixtures of oil/water liquids. *Cellulose* **2012**, *19*, 401–410. [CrossRef]

36. Feng, J.; Nguyen, S.T.; Fan, Z.; Duong, H.M. Advanced fabrication and oil absorption properties of super-hydrophobic recycled cellulose aerogels. *Chem. Eng. J.* **2015**, *270*, 168–175. [CrossRef]

37. Jin, C.; Han, S.; Li, J.; Sun, Q. Fabrication of cellulose-based aerogels from waste newspaper without any pretreatment and their use for absorbents. *Carbohyd. Polym.* **2015**, *123*, 150–156. [CrossRef] [PubMed]

38. Stalder, A.F.; Melchior, T.; Müller, M.; Sage, D.; Blu, T.; Unser, M. Low-bond axisymmetric drop shape analysis for surface tension and contact angle measurements of sessile drops. *Colloid Surf. A Physicochem. Eng. Asp.* **2010**, *364*, 72–81. [CrossRef]

39. Wang, M.; Ma, Y.; Sun, Y.; Hong, S.Y.; Lee, S.K.; Yoon, B.; Chen, L.; Ci, L.; Nam, J.D.; Chen, X.; Suhr, J. Hierarchical porous chitosan sponges as robust and recyclable adsorbents for anionic dye adsorption. *Sci. Rep.* **2017**, *7*, 18054. [CrossRef] [PubMed]

Article

Study of the Suitability of Different Types of Slag and Its Influence on the Quality of Green Grouts Obtained by Partial Replacement of Cement

Francisca Perez-Garcia [1], Maria Eugenia Parron-Rubio [2,*], Jose Manuel Garcia-Manrique [1] and Maria Dolores Rubio-Cintas [2]

[1] Departamento de Ingeniería Civil, Materiales y Fabricación, Universidad de Málaga, 29071 Málaga, Spain; perez@uma.es (F.P.-G.); josegmo@uma.es (J.M.G.-M.)

[2] Departamento de Ingeniería Industrial y Civil, Universidad de Cádiz, 11202 Algeciras, Spain; mariadolores.rubio@uca.es

* Correspondence: mariaeugenia.parron@uca.es

Received: 18 March 2019; Accepted: 8 April 2019; Published: 10 April 2019

Abstract: This paper is part of a research line focused on the reduction of the use of cement in the industry. In this work, the study of work methodologies for the manufacture of green cementitious grout mixtures is studied. Grout is widely used in construction and it requires an important use of raw materials. On the other hand, the steel industry faces the problem of the growing generation of slag wastes due to the increase in steel manufacturing. The green grout aims to achieve the dual objective of reducing the demand for cement and improve the slag waste valorization. Slag is not introduced as an aggregate but through the direct replacement of cement and no additives. The research seeks a product where we can use steel slag intensively, guaranteeing minimum resistance and workability. Results with substitutions between a 25% to 50% and water/cement ratio of 1 are presented. In particular, the suitability of different slags (two Ladle Furnace Slag (LFS) and one Blast Furnace Slag (GGBS)) in the quality of the final product are analyzed. The feasibility of replacing cement with slag and the importance of the origin and pretreatment are highlighted.

Keywords: cementitious grout; green grout; cement; slag substitution; valorization; circular economy

---

1. Introduction

One of the most important challenges that society faces is to achieve a balance between the consumption of raw materials and our need for development. The future evolution of the industrial activity must include criteria of both efficiency and reuse of waste. The dimension of transformation must be limited to a sustainable environment where the global needs could be satisfied without severely compromising those of future generations (Brundlandt Report, 1987) [1]. To achieve this sustainable environment, actions must be taken in the reduction of harmful gas emissions (greenhouse effect) and the reduction of the use of natural raw materials.

One way to reduce the use of raw materials is to lead our efforts toward objectives such as those proposed in the circular economy theory, being aware that we inhabit a world with finite resources.

Therefore, the essence of the circular economy is to optimize the reuse of the generated resources and introduce them back into the production chain. This process, known as waste valorization [2], has become more and more important and today is a field of research with great potential.

Our efforts are focused on the construction industry, in particular, in the process of manufacturing cement-based materials. It is a fundamental element due to its role as a binding component in different mixtures. In order to reach a sustainable environment, the production of cement should be reduced.

One of the reasons is to minimize the extraction of limestone, but more significant is the reduction of energy consumption and on greenhouse gas emissions.

Throughout the entire process, the estimated CO_2 release during clinker manufacture is around 0.7 to 0.9 tons per Portland cement ton, which means that the cement industry generates between 7 and 9% of CO_2 worldwide. The decrease of these emissions is the trigger of the notable research interest in making progress toward reducing the industry's dependence on cement [3–6]. As an alternative to the use of cement, its partial replacement by other materials is proposed. At this point, the idea of valorization of existing waste in the industry itself becomes important.

Steel slag is a byproduct of steel manufacturing, which is obtained by the chemical reactions that take place in the processes of metal formation. It is a complex solution of silicates and oxides produced during the separation of the molten steel from impurities. The properties of the slag produced during steel making depends on many factors, mainly the manufacturing process. According to Setie et al. [2], four types of steel slag can be distinguished: electric arc furnace (EAF) slag, blast furnace slag (GGBS), basic oxygen furnace slag (BOFS), and ladle furnace slag (LFS) [7–9].

The EAF is a strong, dense, nonporous aggregate that is cubical in shape, has good resistance to polishing, and has an excellent affinity to bitumen. Therefore, EAF are more suitable for engineering purposes. The EAF slags can be also divided in two types: oxidizing or black (EAFS) and reducing or white (LFS). In a usual manufacturing process of steel, the EAFS produced is in the order of 110–130 kg per metric ton, and the LFS white one is about 20–30 kg per metric ton.

The increase in steel consumption supposes a proportional increase in the generation of this slag's waste. The strategies of waste valorization of the slags are diverse but not sufficient to achieve in practice a real reuse of these. Either by regulatory or economic problems. The steel slags are used in many areas, from fertilizers to civil industry. In the European Union, it has been used as an additive to make up the cements. The slag properties make them very appropriate for aggressive environments. Slags improve their resistance against salt water and sulphates (maritime facilities). In the last few decades, research efforts have been focused on its use as additives or as aggregate substitutes with arid, both fine and coarse, or as a substitute of arid as a bituminous binder in the pavement layer [10–20].

In the literature, we can find some recent and interesting works about cementitious grouts containing supplementary materials. In 2015, Celik et al. [21] investigate the mixture of rice husk ash in cement-based grout, the rheological properties of the mixture result in the increase in apparent viscosity. Amahjour et al. [22] (2002) or Pastor et al [23] (2016) add fly ash and silica fume to increase mechanical strength.

There are also studies where the substitutions of blast furnace slag are made in small percentages but always with chemical additives. Azadi et al. [24] (2013) worked with chemical additives to optimize the grout. They used sodium silicate (Na_2SiO_3) to increase resistance, sodium carbonate (Na_2CO_3) to reduce bleeding, or triethanolamine (TEA) to promote injection. In 2017, Zhang [25] introduced sodium silicate for quick adjustment.

An interesting research in this area is that done by Krishnamoorthy et al. [26]. In this study a cementitious grout containing supplementary cementitious materials (SCM) (fly ash, silica fume, GGBS) are presented. The results on flow characteristics, strength, and durability show that cementitious grouts containing SCM can be used successfully to repair concrete structures. They reach percentage substitution of 20 to 40% of GGBS slags with water/cement ratios of 0.25 to 0.40.

Huang [27] presented the study on cement ash slurries containing polypropylene (PP), fiber, and super plasticizer. With the addition of PP fiber, better resistance against cracking, sulphate attack, and volume changes was observed, but resulted in a higher viscosity and permeability. Bastien et al. [28] studied the properties of cement slurries with a low ratio of cement and water with superplasticizer and a low proportion of precipitated silica (3% by weight of cement). The rheological properties were also investigated. It is possible to obtain slurry mixtures with zero bleeding, good fluidity, and high compressive strength that meet the requirements for the use of post-tensioning.

Shannag [29] studied cementitious grouts, adding silica fume and natural pozzolan to achieve high performance. This incorporation results in a high fluidity, zero bleeding, high strength, and satisfactory shrinkage.

The literature indicates that the desirable properties for the grouts are that they must possess good fluidity, reduction of bleeding, initial setting time that is not too short, adequate strength, and durability.

Our line of research seeks to advance in the study of the technical feasibility of replacing cement with slag from the steel industry [30], both for the production of green concrete [7,31] and green cementitious grout.

However, there are many factors involved, as the properties of the resulting product related not only mechanical properties, but also durability, that are correlated with the characteristics of the slag used and its proportion.

This paper presents the results obtained in a set of tests for the manufacture of green grouts in a substitution dosage that varies between 0 and 50%. The substitutions have been made on cement Portland (CEM I). Unlike Portland CEM III and IV cements, which already have slag incorporations in their manufacture [32,33], our objective is to evaluate the possibility of working the cementitious grout with recycled material directly [34]. In particular, each mixture has been subjected to a test campaign of a slump test, compressive and bending tests, and exudation test. As mentioned, it is part of a wide "program intended to elaborate a standard to guide the use of steel slag as cement substitute" [7,31].

In the resulting discussion, the optimized simplification of the parameters of the model is considered when admitting that each phase of the material is subject to a similar tension that eliminates the influence of the form, the size, and the disposition of the phases. Therefore, the only factors that this approach considers are the concrete's resistance properties in relation to the replacement of some of its components by slag. The models are complicated, taking into account the case that the cement paste is the connection phase, where the inclusions form a disconnected phase [35].

A fundamental parameter to analyze will be the different behavior obtained (quality of the Green Grout) according to the characteristics of the slag. Two key aspects, the origin of the slag (dependent on the type of steel produced) and the treatments to which the slag have been subjected becomes determinant. In this research, slag from different origins within the country have been studied. The results obtained, such as those presented in a previous work for the case of concrete [7], are very positive and show the feasibility of the mixture.

On the following section, the materials analyzed are described. Then, a brief overview of the tests performed is made. In the next section, results are presented and discussed. Finally, the fundamental conclusion and lines of future research are summarized.

2. Materials

This article focuses on the progress made in replacing material in cementitious grout mixtures. This material has a wide presence in civil constructions. Two of the applications where this research presents potential applicability are the jet grouting, for the improvement of soil, and the execution of deep foundations such as micropiles. In the jet grouting process, the slurry is injected into the soil pores in order to fill and create cohesion, which increases the resistance characteristics or, equivalently, improves its mechanical properties [24,36]. Therefore, the rheological properties of grouts are directly related to the pumping capacity to penetrate holes and cracks [21].

One of the fundamental parameters in cement-based grout is the water/cement ratio by weight. Depending on the application, different relationships are recommended. In applications of foundation by injection, the cementitious grout needs to behave like a fluid able to penetrate the soil or rock so the ratio will vary between 0.6 and 2. In sealed works, it varies between 0.5 and 1. Figure 1 presents a schematic of these recommendations.

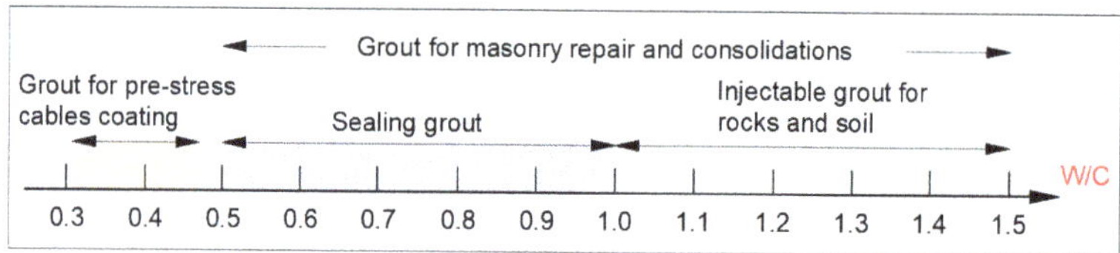

Figure 1. Schematic water–cement ratio (W/C) depending on the application.

Ten different mixes were designed by substituting a 30%, 40% and 50% of cement by slags (% in weight) obtained from three different blast furnaces in Spain, according to previous works with concrete [7,31]. Table 1 shows the cement and slag chemical composition (data provided by the supplying company). Table 2 presents the nomenclature used for the mixtures.

Table 1. Cement and slag chemical composition (data provided by the supplying company).

Slag Origin/Chemical Composition	SiO_2	Al_2O_3	CaO	Fe_2O_3	MgO	Na_2O	K_2O
	%	%	%	%	%	%	%
Cement	20–22	4–10	55.62	4	2	0.3	0.3
Slag GGBS (S1)	35.9	11.2	40	0.3	7.7	0.2	0.4
Slag LFS 1 (S2)	22.28	9.37	56.94	0.84	7.37	0	-
Slag LFS 2 (S3)	15.85	16.53	57.56	0.83	7.7	-	-

Table 2. Cementitious grout mixtures composition.

Mix Denomination	Slag	Substitution
S0	-	0%
S1_30	GGBS	30%
S1_40	GGBS	40%
S1_50	GGBS	50%
S2_30	LFS 1	30%
S2_40	LFS 1	40%
S2_50	LFS 1	50%
S3_30	LFS 2	30%
S3_40	LFS 2	40%
S3_50	LFS 2	50%

The used cement was Portland Cement CEM I 42.5 R (EN 197-1 [37], Holcim, Málaga, Spain). This cement was selected due to the absence of any kind of additive that could mask the results. It was used as reference pattern. Density: 3150 kg/m^3. Specific surface area: >280 m^2/kg. The water–cement ratio used was 1.

Three different slags are used (S1, S2 and S3). The first of them, S1, was a ground granulated blast furnace slag (GGBS) with mechanical processing. It presented a maximum grain size of 0.063 mm, density of 2910 kg/m^3, and specific surface area of 462 m^2/kg. On the other hand, S2 and S3 types were unprocessed ladle furnace slag (LFS) from two different steel mills. The only process they were subjected to was sieving in the lab with a 0.063 mm sieve. The fraction obtained through sieving was 23% and 15% by weight, respectively.

Chemical composition of the slags and cement used in the grouts mixtures is shown in Table 1. This chemical composition was proportionated by the slag supplier companies and it refers to the slags before the sieving process when needed in the laboratory.

As can be seen in Table 1, the chemical composition of the different slags varied for each of them, especially the percentage of SiO_2 and CaO.

The materials employed in this work did not include any type of additive because the objective of the research was to determine how the different slags behaved as substitutes of the binder without being affected by any additional parameter.

3. Tests Description

The mixes described in previous section were subject to different standard tests. The objective of these tests was to evaluate how cement-slag substitution may affect the main properties such as consistency, workability, and mechanical capabilities (compressive and flexural strength).

The batching was prepared according to the European standard used in the manufacture of cement grouts: EN 447 [38]. The laboratory conditions were 24 and 26 °C (temperature) and 40% relative humidity. An electric mixer of robust construction was used, with two speeds of rotation, with a mixing paddle with an anchoring system according to EN 196/1 [39].

Using the dispensing hopper, the Portland cement was incorporated into the mixing bowl with the slag where it was mixed for 90 s, then the water was added, again mixing for 180 s (Figure 2). For each of the mixtures, nine prismatic test specimens of $4 \times 4 \times 16$ cm^3 were prepared. Each one according to the EN 12390-2 standard [40] for hardened concrete where the methods for the manufacture and curing of specimens destined for the performance of resistance tests are described.

Figure 2. Exudation tests.

3.1. Flow Cone Test

The test for the flow of grout mixtures (flow cone method) was determined according to the norm EN 445 [41]. The test determines the time of efflux of a specified volume of fluid cement grout through a standardized flow cone.

Before beginning, the inside of the cone was moistened by filling the cone with water. The water was drained from the cone one minute before the test. The grout was poured slowly to prevent trapped air. The quantity tested was one liter of mix. Once full, the stopwatch was started, and simultaneously, the stopper was removed. The recorded result was be the time in which the entire grout passed through the cone.

3.2. Flexural Strength Test

The test was carried out at 7, 28 and 90 days for all the test pieces (nine prismatic test specimens for each mixture) according to EN 196-1 [39] and EN 196-7 [42] standards for cements. The test machine was equipped with a bending device incorporating two steel support rollers and a third steel loading roller of the same diameter and equidistant from the other two. The length of these rollers was between 45 and 50 mm. The load was applied continuously and without sudden shocks. The force did not begin to be applied until the load roller and the support rollers rest firmly on the specimen. The rate increase R was 16 N/s, according to the expression:

$$R = \frac{2}{3} \times \frac{S \cdot d_1 \cdot d_2^2}{l} \times \frac{N}{s} \tag{1}$$

where d_1 and d_2 are the dimensions of the square section of the specimen and $l = 3 \times d$ is the distance between the rollers in millimeters.

The force signal was provided by an adjustable load cell to the upper, lower, or base bridge. It was formed by a strain gauge Wheatstone bridge, adhered to a structure. The force captured by the load cell was, due to its location, the same as that of the specimen under test; that is, there was a direct coupling between the test piece and the load cell. The machine was controlled by a computer through an ETIWIN control software, with ENAC (National Accreditation Entity in Spain) calibration certificate.

3.3. Compressive Strength Test

The test was performed at 7, 28, and 90 days as a bending test according to standards EN 196-1 [39] and EN 196-7 [42]. The number of specimens tested was 18 and an average value was calculated. According to Neville [43], the compressive strength of the modified cubic specimen would be 5% higher than the standard cubic specimen. An average value was obtained from this study because only two specimens per flexural strength test were performed.

The test was carried out with the same machine, model ETIMATIC-Proetisa H0224 (Production of Technical and Industrial Equipment, Madrid, Spain). The applied pressure was at an invariant rate of 0.5 MPa/s. The compressive strength is given by the expression:

$$f_c = \frac{F}{A_c} \tag{2}$$

where f_c is the compressive strength in MPa, F is the maximum breaking load expressed in N, and A_c is the cross-sectional area of the specimen given in mm^2.

3.4. Exudation Test

This test gives the exudation of the grout. It was carried out according to EN 445 [41] (Figure 2). Exudation was measured as the volume of water remaining on the surface of the mix that was kept protected from evaporation. The variation in volume was measured as a difference in percentage of the volume of the grout between the start and the end of the test. The test mainly measured the volume variation caused by sedimentation or expansion. A transparent tube, approximately 60 mm in internal diameter and around 1 m in length, was used. The tube was placed in a vertical position with the top end open. It ensured a rigid fixation that prevented any movement or vibration. The grout was poured into the tube with a constant flow to ensure that no trapped air remained. The tube was filled to a height, h_o. The ambient temperature of the laboratory was 18.1 °C and the grout acquired a temperature of 18.3 °C. The start time t_0 and the height h_0 were recorded. The height of the cement grout, h_g, was recorded at intervals of 15 min during the first hour, and then at 2 h, 3 h, and 4 h. The height of the exuded water, h_w, was recorded at the same time as the measurements of the grout were made. Possible heterogeneities that could be seen in its appearance through the transparent tube were recorded. The volume variation was:

$$h_w/h_o \times 100\% \tag{3}$$

4. Interfacial Transition Zone Review

Before presenting the data, the basic principles of mixing models based on the interfacial transition zone are reviewed (ITZ) [44]. This model usually applied to concrete assumes that the material is idealized as a composite of mortar and aggregate, where any arbitrarily small volume contains both mortar and aggregate in fixed proportions [45]. The same hypothesis can be applied to the cementitious grout, which properties will be very influenced by the microstructure. This can be classified into three phases: aggregate, cement paste, and the interfacial transition zone (ITZ). ITZ has a critical role. This transition zone has a size comparable with the size of cement grains.

The effects of varying the percentage of slag substitution affects the state of the structure and causes an improvement in fluency. This leads us to expect an increase in the mechanical strength of the hardened grout. There are few studies on this phenomenon, due to it being difficult to find suitable definitions. It is a new diffuse distribution of the particles in the grout. The global grain size does not change, but the permeability evolves. If locally transported particles do not migrate further, an obstruction occurs that accompanies an overpressure. In short, the substitution of cement by slag results in a redistribution of the fine particles without modification of the total solid volume of the specimen.

Below, the formulation of these models are summarized. The partial densities for the three constituents are given by the following expressions:

$$\rho^s = \rho_s(1 - \varphi) \tag{4}$$

$$\rho^w = \rho_w \varphi(1 - c) \tag{5}$$

$$\rho^c = \rho_c \varphi\, c \tag{6}$$

Where ρ_s, ρ_w, and ρ_c are the real densities of the skeleton, water, and fluidized cement paste, respectively. The porosity, φ, defines the proportion of the holes in relation to the total volume. "c" is the concentration in the cement paste of the ITZ zone. It represents the total volume of the particles transported from the skeleton by the filtering forces in the void volume. The fraction of the mass of the fluidized solid is:

$$c_m = c\rho_w / (c\rho_w + (1 - c)\rho_c) \tag{7}$$

Applying the mass conservation equations to each phase, Equation (8) presents the variation of mass for solid phase, Equation (9) for liquid phase, and Equation (10) for fluidized solid:

$$\frac{\partial \rho^s}{\partial t} + \nabla.(\rho^s v^s) = m^s \tag{8}$$

$$\frac{\partial \rho^w}{\partial t} + \nabla.(\rho^w v^w) = 0 \tag{9}$$

$$\frac{\partial \rho^c}{\partial t} + \nabla.(\rho^c v^c) = m^c \tag{10}$$

where "m" represents the typical variation of the mass of that constituent ($m^w = 0$), "t" is the time and ∂ is the gradient operator. Assuming that all the particles transferred from the skeleton re-enter the fluid, the following mass can be established:

$$m^c + m^s = 0 \tag{11}$$

It will be assumed, on the other hand, that the slag transported via filtration in the slurry moves with the same speed as the particles near the ITZ zone. This relationship translates the particular nature of the phenomenon that is considered in the framework of this study. There is, for example, no chemical reactions that cause a divergence between the mass transported from the solid skeleton per unit of time and that which is transformed into cement paste at the same time. The previous hypothesis also assumes that the ITZ does not significantly modify the proportions between the different phases that constitute the initial sample. Therefore, the initial variation and the temporal evolution of ρ_s (density of the solid skeleton) remain negligible. This means, in particular, that this phenomenon does not develop considerably and causes little variation in the initial properties of the fluidized cement paste in, for example, its density.

5. Results and Discussion

In this section, the main results obtained are presented and discussed. The results obtained in each test are summarized for the proposed mixtures (Table 2). In general, the results between them were analyzed based on the reference mixture without any substitution (S0). The results were also compared with the works of Krishnamoorthy et al. [27]. The results presented in this paper had a higher percentage of substitution and a higher W/C ratio in addition to introducing the use of LFS slag. However, the similar evolution of the parameters studied for GGBS slag was checked.

5.1. Flow Cone Test Results

Flow cone test results are shown in Figure 3. It can be observed that there were no significant differences among the mixtures. The flow cone test results were always within an interval from 8.5 to 9 s. The main implication here is that the use of these types of slags as substitute of cement in cementitious grout had no significant effect in the fluidity of the resultant mix, at least in substitution percentages of up to 50%.

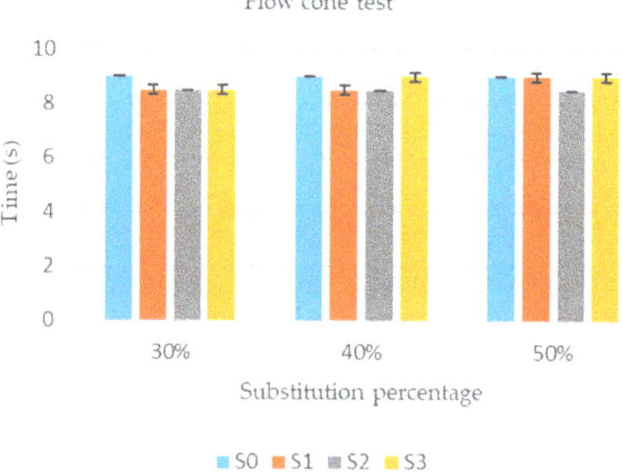

Figure 3. Flow cone test results.

This is an important conclusion because these new mixtures will not present disadvantages in their application with respect to the original ones while being able to take advantage of the same implementation technologies.

5.2. Flexural Strength Test Results

Flexural strength test results are shown in Table 3 for 10 mixtures at 7, 28, and 90 days. In Figure 4, the data are grouped according to time and percentage of substitution for each slag type. Figure 5 also allows for analyzing the behavior of each mixture with respect to the reference. Each result curve (MR) is non-dimensioned with respect to the corresponding value of S0 (MR_{S0}).

None of the mixes with slag substitution achieved the flexural strength reference (S0) at 7 days. However, at 28 days, the MR difference was reduced between S0 and two of the slag types (S1 and S2). One of the effects observed with slag was that the hardening process of the mixture was modified and delayed. As it has been observed in concrete mixtures incorporating GGBS slags as partial cement replacement, the strength at early stages was lower in comparison with traditional concrete. The results obtained in this work are in accordance with the results shown in previous works (Parron-Rubio et al. [7]).

No direct correlations were identified for each type of slag and its improvement with respect to MR.

Table 3. Flexural strength results.

Mixes	7	28	90	% Strength Gain at 90 Days
S0	2.45	3.36	3.92	-
S1_30	1.80	2.65	4.07	3.8%
S1_40	1.70	3.28	4.28	9.18%
S1_50	1.65	3.46	4.63	18.1%
S2_30	1.44	2.90	4.29	9.4%
S2_40	1.70	3.5	3.69	−5.9%
S2_50	1.35	2.61	3.60	−8.2%
S3_30	1.17	2.39	2.21	−43.6%
S3_40	0.88	1.83	1.46	−62.8%
S3_50	0.51	1.16	1.29	−67.1%

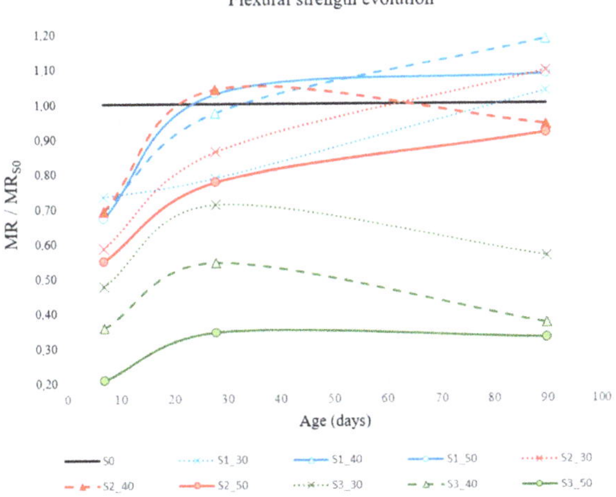

Figure 4. Results of flexural strength test (MR)(MPa).

Figure 5. Flexural strength evolution (MR/MR$_{S0}$).

S1 mixtures showed a higher flexural strength than S2 and S3 mixtures. In the literature, there are some works that indicate that pozzolanic materials with a high SiO_2 content have better mechanical properties than pozzolanic materials with a low content of SiO_2 [7,46]. This can be the reason why S1 mixtures showed the highest flexural strength due to their highest SiO_2 content in comparison with S2 and S3 slags.

It seems that each slag had a particular dosage that optimized its behavior in the test. The S1_40 mix showed the best performance overall in this test, obtaining a flexural strength gain of up to 18.6% at 90 days with the 40% substitution.

S1 slags were unique in presenting an increase in flexural strength at 90 days for every substitution fraction (S1_30, S1_40, and S1_50). On the other hand, S3 slag showed poor results for this test. As it can be observed in Figure 4, flexural strength loss for this type of slag appeared at 7, 28, and 90 days. It seems that in this type of slag, the hardening stopped after 28 days. Actually, our conclusion is that from there it is maintained. We do not consider that the small decrease observed in Figure 4 is representative of any behavior, but is the consequence of some distortion of results.

5.3. Compressive Strength Test Results

Compressive strength test results are shown in Table 4 for 18 test specimens at 7, 28, and 90 days. In Figure 6, the data are grouped according to time and percentage of substitution for each slag type (S1, S2, and S3). Figure 7 also allows for analyzing the behavior of each mixture with respect to the reference. Each result curve (Rs) is non-dimensional with respect to the corresponding value of S0 (Rss_0).

Table 4. Compressive strength results.

Mixes	7	28	90	% Strength Gain at 90 Days
S0	6.29	8.89	12.52	-
S1_30	4.40	6.98	13.25	5.8%
S1_40	4.38	8.66	14.97	19.6%
S1_50	4.50	11.41	16.90	35.0%
S2_30	3.20	6.41	9.27	−26.0%
S2_40	3.98	8.08	9.09	−27.4%
S2_50	3.15	6.27	7.34	−41.4%
S3_30	2.23	4.42	4.42	−64.7%
S3_40	1.65	3.33	3.17	−74.7%
S3_50	1.02	2.05	1.98	−84.2%

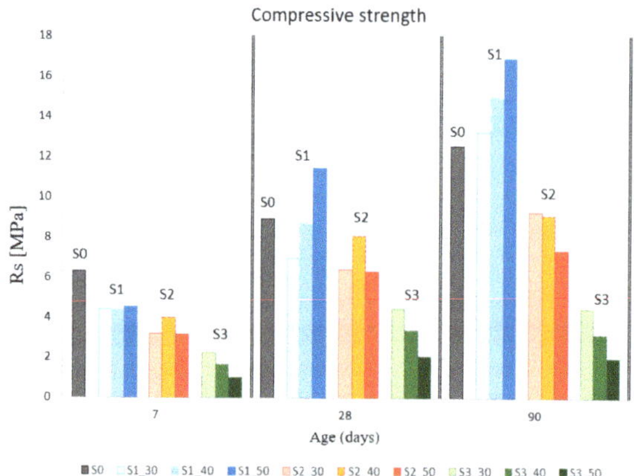

Figure 6. Results of compressive strength test (Rs)(MPa).

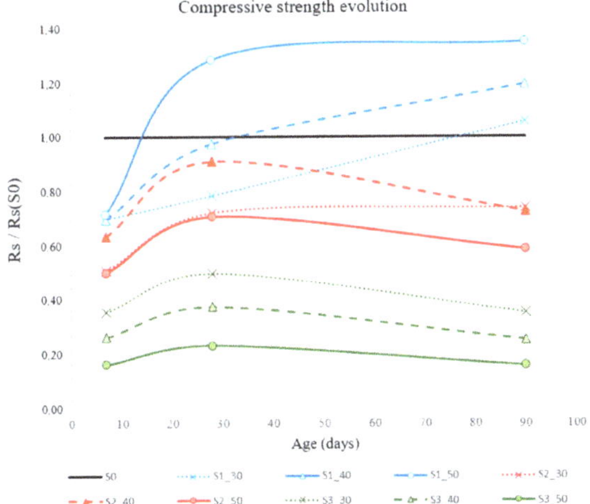

Figure 7. Compressive strength evolution (Rs/Rs($_{S0}$)).

As for the flexural strength results, none of the mixes obtained a compressive strength gain at 7 days with respect to the reference grout mixture. The behavior at 28 and 90 days of the different mixes differed depending on the type of slag and the slag–cement substitution percentage.

The compressive strength of S1 slag grew as the substitution percentage increased. The behavior of the 40% mixture was similar to the 30%. The best compressive performance is attributed to the 50% mixture, which obtained a strength gain at 28 and 90 days of 28.35% and 35%, respectively.

The behavior of S2 slag at 7 days was similar for every substitution percentage and was about 30% less than the reference grout mixture at the same age. S2 mix obtained strength loss at 28 days for every substitution percentage, with the 40% mix (S2_40) the one performing the best, followed by the 30% mix and 50% mix. However, the results for 90 days show a relationship between substitution percentage and compressive strength, with the latter being greater as the substitution percentage decreased.

The compressive strength loss obtained using L3 mixes, at every stage and percentage substitution, was significant. Furthermore, the loss was greater as the substitution percentage increased. As in the case of the flexural strength, the mixtures with GGBS slags showed a higher compressive strength than LFS slags.

Krishnamoorthy et al. [27] presents results of variation of compressive strength with GGBS slags. They test them from porous concrete blocks containing a mixture of supplementary cementitious materials including fly ash, GGBS, and silica fume. One of them was prepared with a grouting mixture of ordinary Portland cement, 40% GGBS as aggregate, water–cement ratio of 0.35, and 1% superplasticizer. Although it is a different product, we can observe a similar slight improvement in the compressive strength.

5.4. Exudation Test Results

The results of the exudation test are shown on the Figure 8.

It can be distinguished in the figure that the mixture without substitution of cement by slag was the one that showed a higher percentage of water exudation at 240 min, obtaining a value of almost 30%.

GGBS (S1) mixtures had low values of exudation at early stages up to 120 min and obtained the lowest percentages at 240 min, except for the S2 mixture with 50% substitution.

LFS1 (S2) slag showed an exudation behavior similar to the reference mixture, although they obtained water exudation percentages superior to those of the rest of the slags.

Regarding the percentage substitution, the results clearly show that with higher substitution percentages, the exudation decreased.

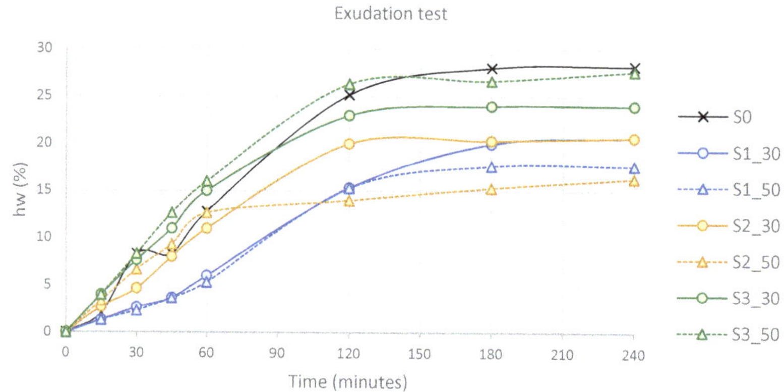

Figure 8. Exudation test.

6. Conclusions

In this paper, experimental results obtained from grouts with cement substitutions by slags in a dosage of up to 50%, W/C ratio of 1, and no additives are presented. The results for different white slags are studied (GGBS and LFS). All the specimens have been tested for exudation, compressive strength, and flexural strength to analyze the feasibility of the mixture for industrial applications. According to the results described in previous sections, the following conclusions can be highlighted:

In general, the mixtures obtained show an improvement in factors such as fluency and viscosity. The slags had a lower density than cement and cause a mixture more fluid. This was an improvement in application where this factor was important.

The mechanical response was less homogeneous and depended greatly on the origin of the slag, as expected. Improvements were observed in the results of compression and bending strength for mixtures with S1 slags (10% in bending and 35% in compression test). However, slag types S2 and S3 gave rise to mixtures with losses of resistance of 85% in compression with respect to the reference. This was due to the lower content of SiO_2 in slags S2 and S3 (LFS) in comparison with S1 slag (GGBS). It has been proven that there was a great difference between slags according to their origin, not only for its composition, but also for the treatment received prior to mixing. Therefore, each generator of slag waste required a study of the goodness of its product in terms of its use as a cementitious substitute. However, the tests seemed to indicate that an adequate treatment increased the potential valorizing of the waste in question.

In the conventional slurry, the aggregate was the least deformed; therefore, this was where the tensions were concentrated. They were then transferred to the ITZ. The green grouts (with cement substitution by slag) had a lower density rate. This increased the speed of the mechanism of the conservation equation of the mass. This filled the microstructural spaces, facilitating the adherence between the aggregate and the cement paste inside the transition.

In addition, the resultant cement grout was a sustainable material with a lower cost in comparison with traditional cement grouts.

The fundamental conclusion of this study was to verify the feasibility of obtaining mixtures with cement substitution by slag. A 50% reduction in the cement used in the mix was achieved, and at the same time, the viability of the grout was maintained. The fluidity obtained allows for use in applications where there can be an intensive use of this grouts, such as jet-grouting for mixtures S1 and S2, or ground improvements for S3 mixtures

Higher cement substitution levels using slag waste may also be possible but this would require further investigation.

GGBS slags improved the mechanical and workability capabilities of the resulting mixture. The LFS slags studied in this work can be employed in other types of works where a high strength is not required. Therefore, this conclusion presents an opportunity to improve waste slag valorization if we progress in the high-level percentage of substitution with slags in ordinary products such as grouts or concrete.

Author Contributions: All the authors conceived and designed the experiments, and analyzed the results; F.P.-G. and M.E.P.-R. performed the experiments; M.D.R.-C. and J.G.-M. coordinated and wrote the paper.

Funding: The authors acknowledge the financial support provided to this work by the European Regional Development Fund (ERDF) as part of the Operational Programme Smart Growth 2014–2020. As well as the Center of Industrial Technological Development (CDTI) of the Ministry of Economy and Competitiveness as a part of the research project IDI-20160509 through the companies DRACE and GEOCISA.

Acknowledgments: The authors acknowledge to the technicians Manuel García Pareja and Maria José Rodríguez Aranda from the Polytechnic School of Algeciras for the preparation of tests and their technical support, and to the doctoral student Alberto Fraile Cava for his contribution to this research.

Conflicts of Interest: The authors declare no conflict of interest.

References

1. Brudtland, G.H. Report of the World Commision on Environement and Development: Our Common Future. Oxford University Press: Oxford, UK, 1987.
2. Setién, J.; Hernández, D.; González, J.J. Characterization of ladle furnace basic slag for use as a construction material. *Constr. Build. Mater.* **2009**, *23*, 1788–1794. [CrossRef]
3. Bilim, C.; Ati, C.D. Alkali activation of mortars containing different replacement levels of ground granulated blast furnace slag. *Constr. Build. Mater.* **2012**, *28*, 708–712. [CrossRef]
4. Hannesson, G.; Kuder, K.; Shogren, R.; Lehman, D. The influence of high volume of fly ash and slag on the compressive strength of self-consolidating concrete. *Constr. Build. Mater.* **2012**, *30*, 161–168. [CrossRef]
5. Özodabaş, A.; Yilmaz, K. Improvement of the performance of alkali activated blast furnace slag mortars with very finely ground pumice. *Constr. Build. Mater.* **2013**, *48*, 26–34. [CrossRef]
6. Rashad, A.M. A comprehensive overview about the influence of different additives on the properties of alkali-activated slag—A guide for Civil Engineer. *Constr. Build. Mater.* **2013**, *47*, 29–55. [CrossRef]
7. Parron-Rubio, M.; Perez-García, F.; Gonzalez-Herrera, A.; Rubio-Cintas, M. Concrete Properties Comparison When Substituting a 25% Cement with Slag from Different Provenances. *Materials* **2018**, *11*, 1029. [CrossRef] [PubMed]
8. Rubio-Cintas, M.D.; Barnett, S.J.; Perez-García, F.; Parron-Rubio, M.E. Mechanical-strength characteristics of concrete made with stainless steel industry wastes as binders. *Constr. Build. Mater.* **2019**, *204*, 675–683. [CrossRef]
9. Gökalp, İ.; Uz, V.E.; Saltan, M.; Tutumluer, E. Technical and environmental evaluation of metallurgical slags as aggregate for sustainable pavement layer applications. *Transp. Geotech.* **2018**, *14*, 61–69. [CrossRef]
10. Rubio Cintas, M.D.; Parrón Vera, M.A.; Contreras de Villa, F. Nuevos usos de las escorias y polvos de humo provocados por la siderurgia. *An. Ing. Mecánica* **2008**, *16*, 1233–1238.
11. Nishigaki, M. Producing permeable blocks and pavement bricks from molten slag. *Stud. Environ. Sci.* **1997**, *71*, 31–40. [CrossRef]
12. Rubio, M.D.; Parrón, M.A.; Contreras, F. Resistencia mecánica de hormigones con sustitución de un porcentaje de cemento por polvos de humo de sílice y escoria de horno de arco eléctrico. In *Comunicaciones V Congreso ACHE*; Asociación Científico-Técnica del Hormigón Estructural: Barcelona, Spain, 2011.
13. Hadjsadok, A.; Kenai, S.; Courard, L.; Michel, F.; Khatib, J. Durability of mortar and concretes containing slag with low hydraulic activity. *Cem. Concr. Compos.* **2012**, *34*, 671–677. [CrossRef]

14. Lam, M.N.T.; Jaritngam, S.; Le, D.H. Roller-compacted concrete pavement made of Electric Arc Furnace slag aggregate: Mix design and mechanical properties. *Constr. Build. Mater.* **2017**, *154*, 482–495. [CrossRef]

15. Xue, Y.; Wu, S.; Hou, H.; Zha, J. Experimental investigation of basic oxygen furnace slag used as aggregate in asphalt mixture. *J. Hazard. Mater.* **2006**, *138*, 261–268. [CrossRef] [PubMed]

16. Behiry, A.E.A.E.M. Evaluation of steel slag and crushed limestone mixtures as subbase material in flexible pavement. *Ain Shams Eng. J.* **2013**, *4*, 43–53. [CrossRef]

17. Sas, W.; Głuchowski, A.; Radziemska, M.; Dzięcioł, J.; Szymański, A. Environmental and geotechnical assessment of the steel slags as a material for road structure. *Materials* **2015**, *8*, 4857–4875. [CrossRef]

18. Mahmoud, E.; Ibrahim, A.; El-Chabib, H.; Patibandla, V.C. Self-Consolidating Concrete Incorporating High Volume of Fly Ash, Slag, and Recycled Asphalt Pavement. *Int. J. Concr. Struct. Mater.* **2013**, *7*, 155–163. [CrossRef]

19. Hybská, H.; Hroncová, E.; Ladomerský, J.; Balco, K.; Mitterpach, J. Ecotoxicity of concretes with granulated slag from gray iron pilot production as filler. *Materials* **2017**, *10*, 505. [CrossRef]

20. Boza, M. Universidad de Holguín Utilización de las escorias de acería como material de construcción. *Ciencia Futuro* **2011**, *1*, 31–40.

21. Celik, F.; Canakci, H. An investigation of rheological properties of cement-based grout mixed with rice husk ash (RHA). *Constr. Build. Mater.* **2015**, *91*, 187–194. [CrossRef]

22. Amahjour, F.; Pardo, P.; Borrachero, M.V. Propiedades De Lechadas De Cemento Fabricadas Con Cementos De Tipo I Y Mezclas Con Cenizas Volantes (Cv)Y Humo De Sílice (Hs). In Proceedings of the VIII Congreso Nacional de Propiedades Mecánicas de Sólidos, Gandia, Spain, 25–28 June 2002; pp. 729–737.

23. Pastor, J.L.; Ortega, J.M.; Flor, M.; López, M.P.; Sánchez, I.; Climent, M.A. Microstructure and durability of fly ash cement grouts for micropiles. *Constr. Build. Mater.* **2016**, *117*, 47–57. [CrossRef]

24. Azadi, M.R.; Taghichian, A.; Taheri, A. Optimization of cement-based grouts using chemical additives. *J. Rock Mech. Geotech. Eng.* **2017**, *9*, 623–637. [CrossRef]

25. Zhang, W.; Li, S.; Wei, J.; Zhang, Q.; Liu, R.; Zhang, X.; Yin, H. Grouting rock fractures with cement and sodium silicate grout. *Carbonates Evaporites* **2018**, *33*, 211–222. [CrossRef]

26. Krishnamoorthy, T.S.; Gopalakrishnan, S.; Balasubramanian, K.; Bharatkumar, B.H.; Rama Mohan Rao, P. Investigations on the cementitious grouts containing supplementary cementitious materials. *Cem. Concr. Res.* **2002**, *32*, 1395–1405. [CrossRef]

27. Huang, W.-H. Improving the properties of cement–fly ash grout using fiber and superplasticizer. *Cem. Concr. Res.* **2001**, *31*, 1033–1041. [CrossRef]

28. Bastien, J.; Dugat, J.; Prat, E. Cement Grout Containing Precipitated Silica and Superplasticizers for Post-Tensioning. *ACI Mater. J.* **1997**, *94*, 291–295. [CrossRef]

29. Shannag, M.J. High-performance cementitious grouts for structural repair. *Cem. Concr. Res.* **2002**, *32*, 803–808. [CrossRef]

30. Kim, J.-H.; Lee, H.-S. Improvement of Early Strength of Cement Mortar Containing Granulated Blast Furnace Slag Using Industrial Byproducts. *Materials* **2017**, *10*, 1050. [CrossRef]

31. Rubio Cintas, M.D.; Parrón Vera, M.A.; Contreras, F.; Rubio, M.D.; Parrón, M.A.; Contreras, F. Method for Producing Cinder Concrete. Patent ES20130000758 20130803, 3 June 2015.

32. Ortega, J.M.; Albaladejo, A.; Pastor, J.L.; Sánchez, I.; Climent, M.A. Influence of using slag cement on the microstructure and durability related properties of cement grouts for micropiles. *Constr. Build. Mater.* **2013**, *38*, 84–93. [CrossRef]

33. Manso, J.M.; Polanco, J.A.; Losañez, M.; González, J.J. Durability of concrete made with EAF slag as aggregate. *Cem. Concr. Compos.* **2006**, *28*, 528–534. [CrossRef]

34. Manso, J.M.; Ortega-López, V.; Polanco, J.A.; Setién, J. The use of ladle furnace slag in soil stabilization. *Constr. Build. Mater.* **2013**, *40*, 126–134. [CrossRef]

35. Mavko, G. Effective Medium Theories. Ph.D. Thesis, Stanford University, Stanford, CA, USA, 2010.

36. Li, S.; Sha, F.; Liu, R.; Zhang, Q.; Li, Z. Investigation on fundamental properties of microfine cement and cement-slag grouts. *Constr. Build. Mater.* **2017**, *153*, 965–974. [CrossRef]

37. *EN 197-1:2011 Cement-Part1: Composition, Specifications and Conformity Criteria for Commons Cements*; European Committee for Standardization: Brussels, Belgium, 14 September 2011.

38. *EN 447: 2009 Grout for Prestressing Tendons. Basic Requirements*; European Committee for Standardization: Brussels, Belgium, 2 September 2009.

39. *EN 196-1:2005 Methods of Testing Cement. Determination of Strength*; European Committee for Standardization: Brussels, Belgium, 22 March 2005.
40. *EN 12390-2:2009 Testing Hardened Concrete Part 2: Making and Curing Specimens for Strength Tests*; European Committee for Standardization: Brussels, Belgium, 30 September 2009.
41. *EN 445: 2011 Grout for Prestressing Tendons*; Test methods; European Committee for Standardization: Brussels, Belgium, 10 May 2011.
42. *EN 196-7:2008 Methods of Testing Cement Part 7: Methods of Taking and Preparing Samples of Cement*; European Committee for Standardization: Brussels, Belgium, 31 January 2018.
43. Neville, A.M. *Properties of Concrete/A.M.Neville*, 4th ed.; Longman: Harlow, UK, 1995.
44. Scrivener, K.L.; Crumbie, A.K.; Laugesen, P. The interfacial transition zone (ITZ) between cement paste and aggregate in concrete. *Interface Sci.* **2004**, *12*, 411–421. [CrossRef]
45. Ortiz, M.; Popov, E.P. Plain concrete as a composite material. *Mech. Mater.* **1982**, *1*, 139–150. [CrossRef]
46. Cánovas, M.F.; Gaitan, V.H. Behavior of steel fibe high strength concrete under impact of projectiles. *Mater. Constr.* **2012**, *62*, 381–396. [CrossRef]

 *materials*

Article

Rapid Immobilization of Simulated Radioactive Soil Waste Using Self-Propagating Synthesized Gd$_2$Ti$_2$O$_7$ Pyrochlore Matrix

Jiali Xue [1], Kuibao Zhang [1,2,*], Zongsheng He [1], Wenwen Zhao [1], Weiwei Li [1], Dayan Xie [1], Baozhu Luo [1], Kai Xu [3] and Haibin Zhang [4,*]

[1] State Key Laboratory of Environment-friendly Energy Materials, Southwest University of Science and Technology, Mianyang 621010, China; xuejiali0304@163.com (J.X.); hezongsheng@swust.edu.cn (Z.H.); zhaowenwen@swust.du.cn (W.Z.); liweiwei@swust.du.cn (W.L.); xiedayan@swust.edu.cn (D.X.); luobaozhu@swust.edu.cn (B.L.)

[2] Sichuan Civil-Military Integration Institute, Mianyang 621010, China

[3] State Key Laboratory of Silicate Materials for Architectures, Wuhan University of Technology, Wuhan 430070, China; kaixu@whut.edu.cn

[4] Institute of Nuclear Physics and Chemistry, China Academy of Engineering Physics, Mianyang 621900, China

* Correspondence: zhangkuibao@swust.edu.cn (K.Z.); hbzhang@caep.cn (H.Z.); Tel.: +86-816-241-9492 (K.Z.)

Received: 22 February 2019; Accepted: 8 April 2019; Published: 10 April 2019

Abstract: A rapid and effective method is necessary in the disposal of severely radioactive contaminated soil waste. Simulated Ce-bearing radioactive soil waste was immobilized by self-propagating high-temperature synthesis (SHS) within 5 min in this study. The main work includes the rapid synthesis of soil waste forms, the analysis of phase composition, microstructure and chemical durability. These results show that the simulated nuclide Ce was successfully immobilized into the pyrochlore-rich waste matrice, whose main phases are SiO$_2$, pyrochlore (Gd$_2$Ti$_2$O$_7$) and Cu. The normalized leaching rates of Si and Na on the 42nd day are 1.86×10^{-3} and 1.63×10^{-2} g·m^{-2}·d^{-1}, respectively. And the normalized leaching rate of Ce also remains at low level (10^{-5}–10^{-6} g·m^{-2}·d^{-1}) within 42 days.

Keywords: Radioactive soil waste; Gd$_2$Ti$_2$O$_7$ pyrochlore; SHS; CeO$_2$; Immobilization

---

1. Introduction

In recent years, nuclear power has been developed rapidly in the world due to its advantages of high efficiency, economy and low carbon emissions. However, the harm caused by the byproduction of nuclear energy, mainly nuclear wastes, can hardly be ignored, especially high-level radioactive waste (HLW) [1]. The radioactive nuclides in HLW, such as ^{137}Cs, ^{90}Sr, ^{239}Pu, ^{235}U, etc., possess the characteristics of long half-life, high toxicity, and high heat generation [2,3]. When these radionuclides enter the soil, the situation becomes more complicated because the presence of soil will increase the cost of disposal [4,5]. Furthermore, soil contaminated by highly radioactive nuclides may pose a long-term threat to organisms due to ecological cycling [6,7].

For radioactive contaminated soil, sand is an inseparable main substance and must be cured together with radionuclides. At present, there are mainly physical landfill and bioremediation methods for the treatment of radioactive contaminated soil. The physical landfill is a time-consuming project, which will destroy the ecosystem of disposal area. At the same time, it may cause further pollution due to leakage during transportation [8]. The bioremediation method requires a long period of restoration, and the growth of plants is limited by climate and geology [9]. In addition to the above methods, vitrification is an effective technology for the immobilization of long half-life wastes, especially for soil wastes contaminated by high radioactive nuclides [10,11]. Particularly, borosilicate glass is the most widely studied and utilized vitrified waste form in the world because of its good radiation resistance, corrosion resistance, and chemical stability [12]. Regrettably, the glass matrice may decompose in geological repositories due to high temperature and high pressure [13,14]. Furthermore, the relatively low thermal stability of vitrified waste forms is also a potential limitation [15]. Compared with glass matrice, ceramic solidified bodies have the advantages of low expansion rate, excellent mechanical properties, and good chemical stability. Subsequently, Synroc has been proposed as a potential alternative host matrix for HLW immobilization based on the isomorphism substitution theory [16–20].

Self-propagating high-temperature synthesis (SHS) is a technology that uses the energy released by exothermic redox reactions to synthesize the final products [21–23]. SHS technology possesses certain technical and cost advantages in the treatment of contaminated soil. According to the characteristics of radionuclides, the composition and proportion of a SHS reaction system can be well designed. In addition, quick pressing (QP) is also introduced to obtain compact samples [24]. Zirconolite-rich matrice and titanate-pyrochlore with excellent chemical durability can also be prepared by SHS/QP [25–30]. SHS/QP technology can synthesize high density ceramic matrix in several minutes, which was considered as a potential method to deal with environmental issues. In this study, $Gd_2Ti_2O_7$ pyrochlore waste matrix was synthesized by SHS for the disposal of simulated radioactive soil waste. Ten wt.% CeO_2 was regarded as a simulate of tetravalent actinide [31]. Silica sand was utilized as the heat insulation material and pressure transfer medium during the SHS/QP process. A series of characterizations were carried out to understand the solidifying mechanism of obtained waste forms. In addition, the aqueous durability was evaluated using the standard Product Consistency Test (PCT) method [32].

2. Materials and Methods

The composition of original soil is listed in Table 1. The soil (200 meshes) and simulated radionuclide Ce^{4+} (CeO_2, Aladdin Industrial Inc., purity $\geq$ 99.99%)) were mixed with the weight ratio of 9:1. The SHS reaction was prepared according to the following chemical equation [28]:

$$4CuO + Gd_2O_3 + 2Ti = Gd_2Ti_2O_7 + 4Cu \tag{1}$$

The raw materials of CuO, Gd_2O_3, and Ti (purity $\geq$ 99.9 wt.%) were purchased from Aladdin Industrial Inc. (Shanghai, China). Different contents of simulated radioactive soil wastes (0 wt.%, 5 wt.%, 10 wt.%, 15 wt.%, 20 wt.%, 25 wt.%) were mixed with the raw materials of SHS reaction (labelled as Cu-0, Cu-5, Cu-10, Cu-15, Cu-20, and Cu-25, respectively). Pretreatment of powder samples is similar as the previous study [28].

Table 1. Soil composition in this study.

Composition	SiO_2	Al_2O_3	Fe_2O_3	CaO	K_2O	MgO	Na_2O	TiO_2
Content (wt.%)	66.32	16.57	5.87	4.67	2.86	1.66	0.81	0.74

The SHS/QP process is shown in Figure 1. The self-propagating combustion reactions were ignited by tungsten wire, which was located at one side with tight contact of the green body and heated by a direct current of about 50 A. The SHS reactants were ignited at high temperature, and the combustion wave automatically propagated to the unreacted region until the reaction's completion. Before densification, the W/Re 5/26 thermocouple was placed in the center of the samples to measure the reaction temperature of Cu-0 to Cu-25 specimens. The unpressurized samples were crushed into fine powders for X-ray diffraction analysis (XRD; X'Pert PRO, PANalytical B.V., Almelo, The Netherlands).

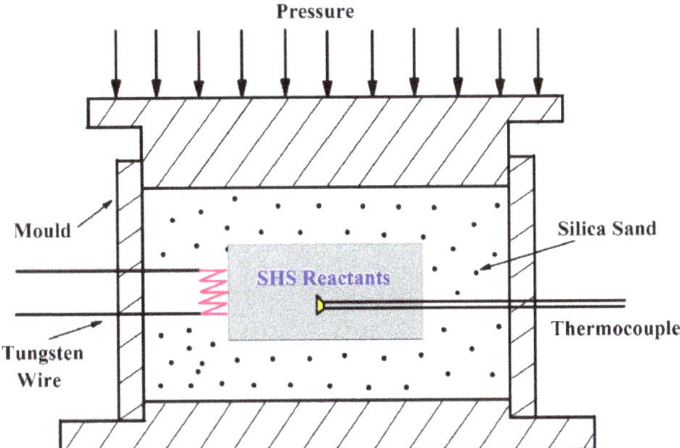

Figure 1. Diagrammatic sketch of the self-propagating high-temperature synthesis/quick pressing (SHS/QP) process.

For densification, the ignited sample was compressed by 50 MPa with 60 s dwelling time after proper combustion delay time. The SHS-ed compact sample was cut and polished to characterize the microstructure and elemental distribution using field-emission scanning electron microscopy (FESEM; Zeiss Ultra-55, Oberkochen, Germany) and energy-dispersive X-ray spectroscopy (EDX, ULTRA 55, ZEISS, Oberkochen, Germany). The chemical durability of waste form was tested by the Product Consistency Test (PCT) standard. The concentrations of Na and Si in leachate were determined by inductively coupled plasma (ICP) analysis (iCPA 6500, ThermoFisher, Waltham, MA, USA), while that of Ce was obtained by inductively coupled plasma-mass spectrometry (ICP-MS) analysis using an Agilent 7700× spectrometer (Agilent, Santa Clara, CA, USA). The normalized release rates were calculated as the following formula:

$$NR_i = \frac{C_i \cdot V}{f_i \cdot S_A \cdot t} \tag{2}$$

where C_i is the concentration of element i in the solution, V is the volume of the leachate (m^3), S_A is the surface area of powder specimen (m^2), f_i is the mass fraction of element i in the sample (wt.%) and t is the leaching duration (d). The S_A/V ratio is about 2000 m^{-1}, which is derived from the standard test method for The Product Consistency Test (ASTM c 1285-02) [32]. In this standard, the waste particles are assumed to be spherical and the average particle diameter is 1.12×10^{-4} m for -100 (0.149 mm) to $+200$ (0.074 mm) meshes particles. Therefore, the average particle area and volume are calculated as 3.90×10^{-8} m^2 and 7.25×10^{-13} m^3, respectively. The average particle mass is calculated to be 1.96×10^{-6} g. Thus, there are 1 g/1.96×10^{-6} g = 5.11×10^5 particles in 1 g powder waste form with -100 to $+200$ meshes particles. Thus, the total surface area of 1 g powder with -100 to $+200$ meshes particles is calculated to be 1.99×10^{-2} m^2. As long as the density and particle size of waste form remain comparable during the leaching tests, this parameter will remain at a constant value and doesn't need to be calculated every time.

3. Results and Discussion

3.1. Temperature and Powder XRD Analysis

The combustion process of the designed SHS reaction takes about 10 s after tungsten wire ignition. The center temperature of all samples in SHS reactions are measured and depicted in Figure 2. With the increment of soil wastes, the center temperatures of Cu-0, Cu-5, Cu-10, Cu-15, Cu-20, and Cu-25 samples decrease from 1679 to 1052 °C in Figure 2. The center temperature of the Cu-0 sample is the highest at 1679 °C, while the Cu-25 sample with the maximum soil content exhibits the lowest temperature at 1052 °C. Apparently, the increase of soil wastes led to the decrease of SHS reaction temperature.

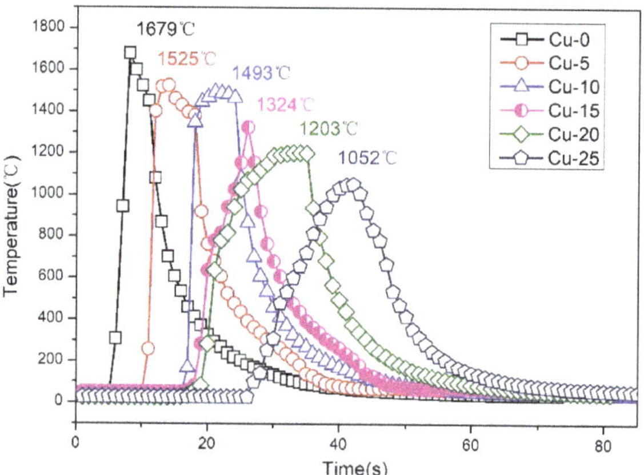

Figure 2. Real temperature curves of all samples during SHS reaction.

The XRD patterns in Figure 3 show that the specimens with soil waste (Cu-5 to Cu-25 samples) are composed of Cu, SiO_2, and $Gd_2Ti_2O_7$ (PDF No. 23-0259), while the sample without soil waste only contains $Gd_2Ti_2O_7$ pyrochlore. From Cu-5 to Cu-25 specimens, the main phase of all samples is $Gd_2Ti_2O_7$ pyrochlore, demonstrating that the increase of soil wastes does not change the phase composition. In Figure 3, the content of Cu in these SHS-ed samples increases with the increment of soil content, but Cu is hardly found in the Cu-0 sample. Because the temperature of Cu-0 reaction is the highest, the Cu melts and condenses into bulky grains during the high temperature reaction. With the decrease of reaction temperature, the size of copper particles decreases. Meanwhile, all SHS-ed samples were ground into powder for XRD testing, where the granulated Cu was sifted out directly. By contrast, the content change of SiO_2 has no regular pattern, which may be affected by the heat insulator silica sand. However, unknown phases appear in the Cu-25 sample, which may be related with the large amount of simulated radioactive soil. Therefore, the Cu-20 specimen was selected for further analysis.

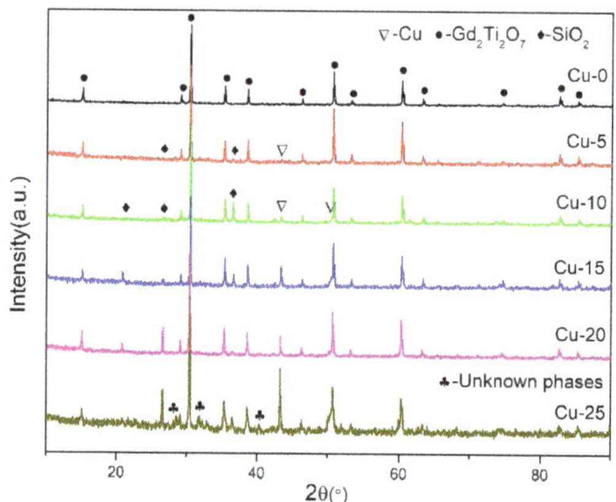

Figure 3. X-ray diffraction (XRD) patterns of all SHS-ed samples.

3.2. Raman Analysis and Microstructure Characterization

Raman spectroscopy was carried out to further analyze the crystal structure and internal bonds of pyrochlore. Raman spectroscopy is an important technique, especially in systems where oxygen displacement induces structure transformation, such as distinguishing fluorite from pyrochlore in pyrochlore ceramics [33]. Different from the $A_2B_2O_7$ fluorite structure with only one F_{2g} vibration mode, the pyrochlore structure contains six Raman modes (A_{1g}, E_g, and $4F_{2g}$). Typical wavenumbers of pyrochlore phase at room temperature are 520 cm^{-1} (A_{1g}), 330 cm^{-1} (E_g), and 200, 310, 450, 580 cm^{-1} ($4F_{2g}$) [33,34]. For Ti-pyrochlore, the most prominent characteristic of Raman spectra are the intensive band at 320 cm^{-1} and the A_{1g} band at 520 cm^{-1}. The band around 320 cm^{-1} includes $E_g + F_{2g}$ modes with very close frequency, which is mostly attributed to O–A–O bond vibration. The A_{1g} band at 520 cm^{-1} is believed to be related to A–O stretching [35,36].

The Raman spectra of Cu-0, Cu-10, and Cu-20 samples are shown in Figure 4. The six Raman active vibration modes (A_{1g}, E_g and $4F_{2g}$) are explicitly assigned. In addition, the Si–O stretching vibration at 1100 cm^{-1} and the Si–O–Si symmetric bending vibration near 700 cm^{-1} are also included. The Raman spectra peaks of three specimens are similar except for some changes in strength, which means the pyrochlore structure of $Gd_2Ti_2O_7$ remains unchanged. In particular, the characteristic F_{2g} (200 cm^{-1} and 455 cm^{-1}) bands are well defined in the Cu-0 specimen. On the contrary, the vibration intensity of $E_g + F_{2g}$ modes (320 cm^{-1}) and A_{1g} mode (520 cm^{-1}) increase significantly in the Cu-10 and Cu-20 samples. It is evident that this drastic change is due to the addition of simulated radioactive soil. On the basis of previous literatures [33–36], we preliminarily speculate that some ions in the simulated radioactive soil (possibly containing Ce) occupy the A and B sites of pyrochlore structure, resulting in steep changes of oxygen ions' environment and peak intensity of Raman spectra.

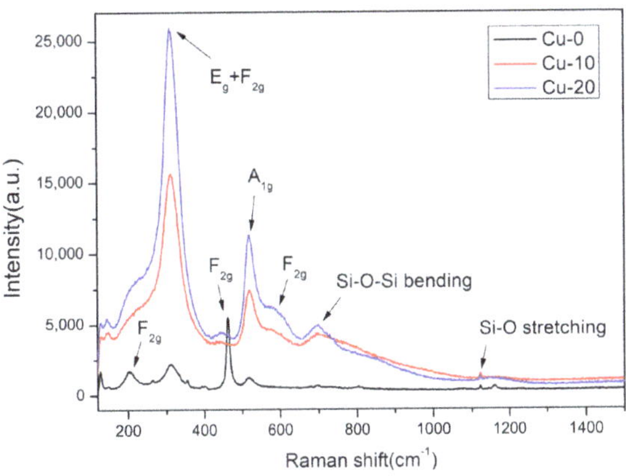

Figure 4. Raman spectra of the Cu-0, Cu-10, and Cu-20 samples.

As shown in Figure 5, the microstructure and elemental distribution of the compact Cu-20 specimen are exhibited in the SEM and elemental mapping images. It can be found that the pores mainly exist in the ceramic matrix rather than the copper phase. It may be argued that the melting point of copper (1083.4 °C) is lower than the combustion temperature of the Cu-20 sample. Therefore, gas can easily be discharged from the copper into the ceramic matrix. The Cu-20 sample consists of four phases, labeled as A, B, C, D in Figure 5a. According to Figure 5b–f and XRD analysis, we speculate that the A region is copper, the B region should be $Gd_2Ti_2O_7$, the C region is SiO_2, and the D region represents TiO_2. The impurity TiO_2 phase is produced from the raw materials of the reaction system and the original soil. However, no TiO_2 exists in the previous XRD result of Cu-20. It is possible that the diffraction peaks of TiO_2 are not obvious because of its low content.

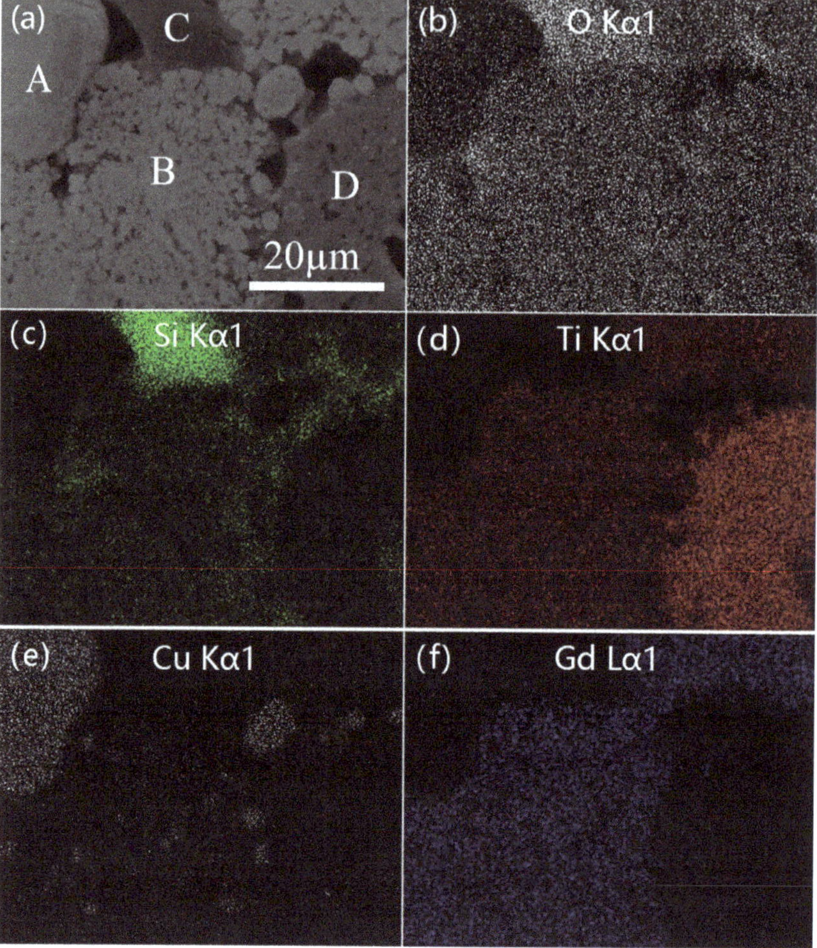

Figure 5. (**a**) SEM image of Cu-20 specimen, and element mapping images of (**b**) O, (**c**) Si, (**d**) Ti, (**e**) Cu, (**f**) Gd.

The EDX elemental spotting analysis of the Cu-20 sample is presented in Figure 6. The EDX spotting image of "B" phase in Figure 6a is presented in Figure 6b. Combined with the XRD and EDX mapping results, the existence of Gd, Ti, Ce, and O in the EDX spotting spectra indicates that the "B" phase is Ce doped $Gd_2Ti_2O_7$ pyrochlore phase. The average elemental quantities are acquired by taking at least five points of "A" area as listed in the inserted table of Figure 6b, which results in the chemical formulation of $Gd_{1.96}Ti_{1.94}Ce_{0.09}O_7$. Meanwhile, a small amount of Ce is also found in the soil phase according to Figure 6c, indicating that the simulated nuclide Ce of radioactive soil waste can exist in both the pyrochlore phase and soil phase. At the same time, most of the elements in original soil are retained in the soil phase. Figure 6d shows that only Ti and O are present in the D region, which confirms that the D phase is TiO_2.

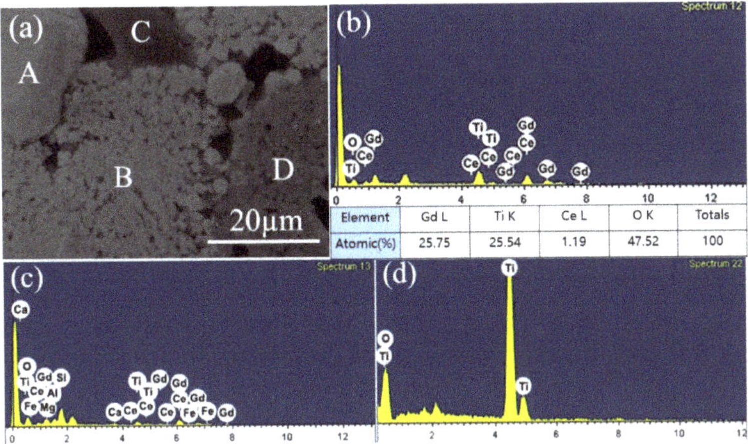

Element	Gd L	Ti K	Ce L	O K	Totals
Atomic(%)	25.75	25.54	1.19	47.52	100

Figure 6. Energy dispersive X-ray spectroscopy (EDX) elemental spotting analysis: (**a**) Representative SEM image of the Cu-20 sample, (**b**) EDX spectrum and elemental composition of the labeled "B" area in (**a**), (**c**) EDX spectrum of region C in (**a**), (**d**) EDX spectrum of region D in (**a**).

3.3. Chemical Durability Measurement by PCT Leaching Test

The leaching performance of nuclear waste forms is a significant indicator for estimating the chemical stability [37]. The 1–42 days normalized elemental leaching rates of Si, Na, and Ce of the Cu-20 sample are depicted in Figure 7. With the extension of soaking time, the normalized leaching rates of Na and Si show a downward trend from 1 to 14 days. Then, NR_{Na} and NR_{Si} exhibit slight ascension within 14–28 days, and finally both of them reach the lowest values in 42 days. The day 1 and day 42 NR_{Si} values are 2.04×10^{-2} and 1.86×10^{-3} g·m^{-2}·d^{-1}, which represent excellent stability. The lowest value of NR_{Na} is 1.63×10^{-2} g·m^{-2}·d^{-1} after 42 days of leaching. This rule is not applicable for Ce, whose leaching rate has no definite regulation. The NR_{Ce} varies from 7.20×10^{-5} g·m^{-2}·d^{-1} of 42 days to 2.20×10^{-6} g·m^{-2}·d^{-1} of 7 days, which is about five orders of magnitude lower than the NR_{Na} values. The NR_{ce} values vary irregularly but remain at a low level (10^{-5}–10^{-6} g·m^{-2}·d^{-1}). Compared with our previous research results [38], the leaching rate of Si is slightly lower in this study. The NR_{Na} of the Cu-20 sample is similar to that of borosilicate glass-ceramics [39]. Nevertheless, the leaching performance of simulated nuclide in Cu-20 sample is significantly lower than that of typical vitreous products [40].

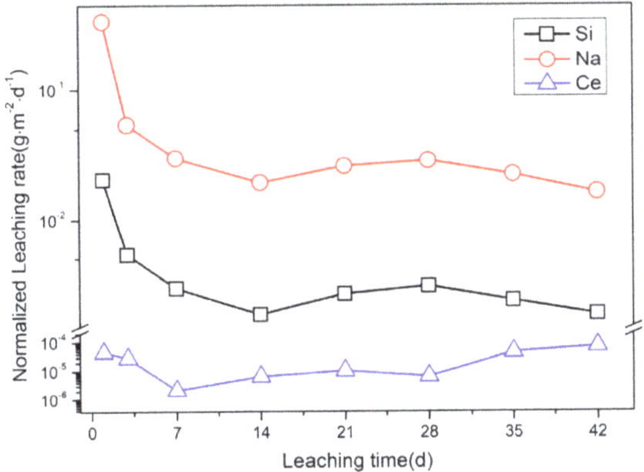

Figure 7. Normalized leaching rates of Si, Na, and Ce from day 1–42.

4. Conclusions

In summary, a series of $Gd_2Ti_2O_7$-based waste forms containing 5–25 wt.% simulated radioactive contaminated soil have been successfully synthesized by SHS in 5 min. The obtained products are multiphase composite materials composed of SiO_2, $Gd_2Ti_2O_7$, and Cu. Furthermore, the simulated nuclide Ce exists in pyrochlore and soil phases simultaneously, which indicates that Ce migrates partly from soil to pyrochlore phase during the SHS reaction. The solidified body of Cu-20 sample exhibits high stability. The 42 days NR_{Si} and NR_{Na} are as low as 1.86×10^{-3} and 1.63×10^{-2} $g \cdot m^{-2} \cdot d^{-1}$, respectively. And the 1–42 days NR_{Ce} values also remain at a low level (10^{-5}–10^{-6} $g \cdot m^{-2} \cdot d^{-1}$). Based on the analysis of phase composition, microstructure, and chemical durability, the application potential of SHS technology in the rapid disposal of radioactive soil wastes is revealed.

Author Contributions: Conceptualization, K.Z., B.L. and H.Z.; Methodology, K.Z. and J.X.; Formal Analysis, J.X., K.X., Z.H. and W.Z.; Investigation, J.X., W.L., D.X. and Z.H.; Writing—Original Draft Preparation, J.X.; Writing—Review & Editing, K.Z. and B.L.

Funding: This research was funded by the National Natural Science Foundation of China (No. 51672228), the Project of State Key Laboratory of Environment-friendly Energy Materials (Southwest University of Science and Technology, No. 16kffk05 and 17FKSY0104) and Science Development Foundation of China Academy of Engineering Physics.

Conflicts of Interest: The authors declare no conflict of interest.

References

1. International Atomic Energy Agency. *Design and Operation of High Level Waste, Vitrification and Storage Facility (Technical Report Series No. 176)*; IAEA: Vienna, Austria, 1977.

2. Ojovan, M.I.; Lee, W.E. *An Introduction to Nuclear Waste Immobilization*; Elsevier Ltd.: Oxford, UK, 2005; pp. 213–267.

3. Caurant, D.; Loiseau, P.; Majérus, O.; Aubin-Chevaldonnet, V.; Bardez, I.; Quintas, A. *Glasses, Glass-Ceramics and Ceramics for Immobilization of Highly Radioactive Nuclear Wastes*; Nova Science Publishers: New York, NY, USA, 2009.

4. Zhu, Y.G.; Shaw, G. Soil contamination with radionuclides and potential remediation. *Chemosphere* **2000**, *41*, 121–128. [PubMed]

5. Mao, X.H.; Qin, Z.G.; Yuan, X.N.; Wang, C.M.; Cai, X.N.; Zhao, W.X.; Zhao, K.; Yang, P.; Fan, X.L. Immobilization of simulated radioactive soil waste containing cerium by self-propagating high-temperature synthesis. *J. Nucl. Mater.* **2013**, *443*, 428–431. [CrossRef]

6. Gavrilescu, M.; Pavel, L.V.; Cretescu, I. Characterization and remediation of soils contaminated with uranium. *J. Hazard. Mater.* **2009**, *163*, 475–510. [CrossRef] [PubMed]

7. Yasunari, T.J.; Stohl, A.; Hayano, R.S.; Burkhart, J.F.; Eckhardt, S.; Yasunari, T. Cesium-137 deposition and contamination of Japanese soils due to the Fukushima nuclear accident. *Proc. Natl. Acad. Sci. USA* **2011**, *108*, 19530–19534. [CrossRef] [PubMed]

8. Xie, Y.; Wu, T.; Shi, Z.K.; Zhang, D. A very low level radioactive waste landfill soil characteristics and properties of strontium block. *Chem. Res. Appl.* **2013**, *25*, 558–562.

9. Lloyd, J.R.; Renshaw, J.C. Bioremediation of radioactive waste: Radionuclide-microbe interactions in laboratory and field-scale studies. *Curr. Opin. Biotech.* **2005**, *16*, 254–260. [CrossRef]

10. Zhang, S.; Ding, Y.; Lu, X.R.; Mao, X.L.; Song, M.X. Rapid and efficient disposal of radioactive contaminated soil using microwave sintering method. *Mater. Lett.* **2016**, *175*, 165–168.

11. Zhang, S.; Shu, X.Y.; Chen, S.Z.; Yang, H.M.; Hou, C.X.; Mao, X.L.; Chi, F.T.; Song, M.X.; Lu, X.R. Rapid immobilization of simulated radioactive soil waste by microwave sintering. *J. Hazard. Mater.* **2017**, *337*, 20–26. [PubMed]

12. Gin, S.; Abdelouas, A.; Criscenti, L.J.; Ebert, W.L. An international initiative on long-term behavior of high-level nuclear waste glass. *Mater. Today* **2013**, *16*, 243–248. [CrossRef]

13. Mccarthy, G.J.; Ebert, W.L.; Roy, R.; Scheetz, B.E. Interactions between nuclear waste and surrounding rock. *Nature* **1978**, *273*, 216–217. [CrossRef]

14. Loiseau, P.; Caurant, D.; Majerus, O.; Baffier, N.; Fillet, C. Crystallization study of (TiO$_2$, ZrO$_2$)-rich SiO$_2$-Al$_2$O$_3$-CaO glasses. Part II. Surface and internal crystallization processes investigated by differential thermal analysis (DTA). *J. Mater. Sci.* **2003**, *38*, 843–852.

15. Caurant, D.; Majérus, O.; Loiseau, P.; Bardez, I.; Baffier, N.; Dussossoy, J.L. Crystallization of neodymium-rich phases in silicate glasses developed for nuclear waste immobilization. *J. Nucl. Mater.* **2006**, *354*, 143–162.

16. Ringwood, A.E.; Kesson, S.E.; Ware, N.G.; Hibberson, W.; Major, A. Immobilization of high level nuclear reactor wastes in SYNROC. *Nature* **1979**, *278*, 219–223. [CrossRef]

17. Franck, P.; John, M.H.; Urs, S. The current state and future of accessory mineral research. *Chem. Geol.* **2002**, *191*, 3–24.

18. Vance, E.R. Development of ceramic waste forms for high-level nuclear waste over the last 30 years. *Mater. Res. Soc. Symp. Proc.* **2006**, *985*, 135–140. [CrossRef]

19. Weber, W.J.; Navrotsky, A.; Stefanovsky, S.; Vance, E.R.; Vernaz, E. Materials science of high-level nuclear waste immobilization. *MRS Bull.* **2009**, *34*, 46–53. [CrossRef]

20. Zhang, K.B.; Yin, D.; Han, P.W.; Zhang, H.B. Two-step synthesis of zirconolite-rich ceramic waste matrice and its physicochemical properties. *Int. J. Appl. Ceram. Technol.* **2018**, *15*, 171–178. [CrossRef]

21. Muthuraman, M.; Dhas, N.A.; Patil, K.C. Combustion synthesis of oxide materials for nuclear waste immobilization. *Bull. Mater. Sci.* **1994**, *17*, 977–987. [CrossRef]

22. Muthuraman, M.; Patil, K.C.; Senbagaraman, S.; Umarji, A.M. Sintering, microstructure and dilatometric studies of combustion synthesized Synroc phases. *Mater. Res. Bull.* **1996**, *31*, 1375–1381. [CrossRef]

23. He, Z.S.; Zhang, K.B.; Xue, J.L.; Zhao, W.W.; Zhang, H.B. Self-propagating chemical furnace synthesis of nanograin Gd$_2$Zr$_2$O$_7$ ceramic and its aqueous durability. *J. Nucl. Mater.* **2018**, *512*, 385–390. [CrossRef]

24. Zhang, K.B.; Wen, G.J.; Zhang, H.B.; Teng, Y.C. Self-propagating high-temperature synthesis of Y$_2$Ti$_2$O$_7$ pyrochlore and its aqueous durability. *J. Nucl. Mater.* **2015**, *465*, 1–5.

25. Zhang, K.B.; He, Z.S.; Peng, L.; Zhang, H.B.; Lu, X.R. Self-propagating synthesis of Y$_{2-x}$Nd$_x$Ti$_2$O$_7$ pyrochlore and its aqueous durability as nuclear waste form. *Scripta. Mater.* **2018**, *146*, 300–303.

26. Zhang, K.B.; He, Z.S.; Xue, J.L.; Zhao, W.W.; Zhang, H.B. Self-propagating synthesis of Y$_{2-x}$Nd$_x$Ti$_2$O$_7$ pyrochlores using CuO as the oxidant and its characterizations as waste form. *J. Nucl. Mater.* **2018**, *507*, 93–100. [CrossRef]

27. Zhang, K.B.; Yin, D.; Peng, L.; Wu, J.J. Self-propagating synthesis and CeO$_2$ immobilization of zirconolite-rich composites using CuO as the oxidant. *Ceram. Int.* **2017**, *43*, 1415–1423. [CrossRef]

28. Peng, L.; Zhang, K.B.; Yin, D.; Wu, J.J.; He, S.H.; He, H.M. Self-propagating synthesis, mechanical property and aqueous durability of Gd$_2$Ti$_2$O$_7$ pyrochlore. *Ceram. Int.* **2016**, *42*, 18907–18913. [CrossRef]

29. Peng, L.; Zhang, K.B.; He, Z.S.; Yin, D.; Xue, J.L.; Xu, C.; Zhang, H.B. Self-propagating high-temperature synthesis of ZrO$_2$ incorporated Gd$_2$Ti$_2$O$_7$ pyrochlore. *J. Adv. Ceram.* **2018**, *7*, 41–49. [CrossRef]

30. He, Z.S.; Zhang, K.B.; Peng, L.; Zhao, W.W.; Xue, J.L.; Zhang, H.B. Self-propagating plus quick pressing synthesis and characterizations of Gd$_{2-x}$Nd$_x$Ti$_{1.3}$Zr$_{0.7}$O$_7$ ($0 \leq x \leq 1.4$) pyrochlores. *J. Nucl. Mater.* **2018**, *504*, 61–67. [CrossRef]

31. Kim, H.S.; Joung, C.Y.; Lee, B.H.; Oh, J.Y.; Koo, Y.H.; Heimgartner, P. Applicability of CeO$_2$ as a surrogate for PuO$_2$ in a MOX fuel development. *J. Nucl. Mater.* **2008**, *378*, 98–104. [CrossRef]

32. ASTM Committee. *Standard Test Methods for Determining Chemical Durability of Nuclear, Hazardous, and Mixed Waste Glasses and Multiphase Glass Ceramics: The Product Consistency Test (PCT)*; ASTM International: West Conshohocken, PA, USA, 2002.

33. Mączka, M.; Hanuza, J.; Hermanowicz, K.; Fuentes, A.F.; Matsuhira, K.; Hiroi, Z. Temperature-dependent Raman scattering studies of the geometrically frustrated pyrochlores Dy$_2$Ti$_2$O$_7$, Gd$_2$Ti$_2$O$_7$ and Er$_2$Ti$_2$O$_7$. *J. Raman Spectrosc.* **2008**, *39*, 537–544. [CrossRef]

34. Mączka, M.; Sanjuán, M.L.; Fuentes, A.F.; Macalik, L.; Hanuza, J.; Matsuhira, K.; Hiroi, Z. Temperature-dependent studies of the geometrically frustrated pyrochlores Ho$_2$Ti$_2$O$_7$ and Dy$_2$Ti$_2$O$_7$. *Phys. Rev. B* **2009**, *79*, 214437. [CrossRef]

35. Zhang, F.X.; Manoun, B.; Saxena, S.K. Pressure-induced order-disorder transitions in pyrochlore RE$_2$Ti$_2$O$_7$ (RE = Y, Gd). *Mater. Lett.* **2006**, *60*, 2773–2776. [CrossRef]

36. Sanjuán, M.L.; Guglieri, C.; DíazMoreno, S.; Aquilanti, G.; Fuentes, A.F.; Olivi, L.; Chaboy, J. Raman and X-ray absorption spectroscopy study of the phase evolution induced by mechanical milling and thermal treatments in R$_2$Ti$_2$O$_7$ pyrochlores. *Phys. Rev. B* **2011**, *84*, 104207. [CrossRef]

37. Zhang, Y.; Stewart, M.W.A.; Li, H.; Carter, M.L.; Vance, E.R.; Moricca, S. Zirconolit-rich titanate ceramics for immobilisation of actinides-Waste form/HIP can interactions and chemical durability. *J. Nucl. Mater.* **2009**, *395*, 69–74. [CrossRef]
38. Xue, J.L.; Zhang, K.B.; He, Z.H.; Zhao, W.W.; Li, W.W.; Xie, D.Y.; Zhang, H.B. Rapid disposal of simulated Ce-bearing radioactive soil waste using self-propagating synthesized zirconolite-rich waste matrice. *Ceram. Int.* **2018**, *44*, 14534–14540.
39. Wu, L.; Li, Y.X.; Teng, Y.C.; Meng, G.L. Preparation and characterization of borosilicate glass-ceramics containing zirconolite and titanite crystalline phases. *J. Non-Cryst. Solids* **2013**, *380*, 123–127. [CrossRef]
40. Ojovan, M.I.; Lee, W.E. Glassy Wasteforms for Nuclear Waste Immobilization. *Metall. Mater. Trans. A* **2011**, *42*, 837–851. [CrossRef]